The Revolt of The Star Men

by

RAYMOND Z. GALLUN

The Revolt of The Star Men
by Raymond Z. Gallun

ISBN: 978-93-59321-78-3

Published by

DOUBLE 9 BOOKS

2/13-B, Ansari Road, Daryaganj
New Delhi – 110002
info@double9books.com
www.double9books.com
Tel. 011-40042856

ABOUT THE AUTHOR

Raymond Zinke Gallun (March 22, 1911 – April 2, 1994) was a science fiction author from the United States. Gallun (rhymes with "balloon") was the son of Adolph and Martha Zinke Gallun and was born in Beaver Dam, Wisconsin. In 1928, he graduated from high school. He dropped out of college after a year and traveled through Europe, living as a vagabond and doing a variety of jobs all over the world in the years leading up to World War II. At the age of 16, Gallun penned his first two stories, "The Space Dwellers" and "The Crystal Ray" (both published in 1929). He was a member of the early sci-fi pulp writers that popularized the genre. In the 1930s, he sold numerous popular stories to pulp publications. His first notable story was "Old Faithful" (1934). "The Gentle Brain" appeared in "Science Fiction Quarterly" under the pen name Arthur Allport. People Minus X, his debut book, was released by Simon & Schuster in 1957, followed by The Planet Strappers (Pyramid) in 1961. The Best of Raymond Z. Gallun, published by Ballantine in 1978, is a selection of his early work. Gallun received the I-CON Lifetime Achievement Award at I-CON IV in 1985; the award was eventually renamed The Raymond Z. Gallun Award.

CONTENTS

It was in the reading room of the Neilson-Aldebar space liner, *Ekova*, that two young people came unexpectedly upon a third person who sat alone, absently skimming through a copy of the *Interplanetarian*. When the girl caught sight of him she uttered a little acclamation of surprise. "Hekki-you!" she cried.

The one addressed looked up. A smile of greeting came over his swarthy, aristocratic features. "Hello, Jan. It is I—none other," he said. "Aren't you glad to see me?" Here he shot a quick glance at the girl's companion.

"Why certainly I am, Hekki," she replied a trifle nervously. "But how can it be? A week ago you left for the deepest, most mysterious part of the Taraal desert on Mars, to collect objects of ancient art, and now you are here. Where have you kept yourself during the voyage?"

The other smiled again—this time a cryptic, secretive smile. "Business," he said mysteriously. "It called me to Earth at the last moment, and since we left the docks at Taboor, it has kept me occupied in my stateroom. This is but the third time I have ventured out of it. Alka brought me my meals." Hekki arched his finely penciled eyebrows slightly as he looked up at the lady's companion. "And you too have had business, Janice," he added. "A new boy friend?" There was a hint of something unpleasant in his tone, but the girl ignored it.

She nodded her golden head. "We met on the night of the departure from Mars, and since then, we've had a happy week together. Austin," she said, turning to the youth, "I want you to know Hekalu Selba of Taboor. Hekki, this is Austin Shelby, who hails from Chicago. You ought to get along well together, because you are both so interested in mechanics," she added. The men shook hands. For the past few moments Shelby had been trying to analyze from the scanty data at hand the character of Hekki. He saw the tapering, effeminate hands—one twiddled nervously a long Martian cigarette—the dark straight hair and fine features; the mouth, that could curl so insolently; the faultless, white silk clothing.

Shelby decided that he did not like Hekki. The reason at first seemed obvious, but presently the young Earthman realized that his feeling towards this child of the Red Planet was stronger than mere dislike. What was the explanation? Was it because Hekki was a friend of Janice Darell? Since he had met her aboard the *Ekova* on this glorious return to Earth, after having spent a whole Martian year at an engineering school at Taboor, Shelby had learned to know love. Was he jealous of this noble of another world? A little, perhaps. But this did not account for the vague, sinister aura he sensed about Hekalu Selba.

Something in Shelby's brain was trying to surge its way to the surface of his consciousness; he struggled with it, and it came out clear. Only thirty-six hours before, during the period designated for sleep, he had wandered into a seldom frequented passageway, high up in the hull of the *Ekova*. Here there were portholes through which he could see the curving metal expanse of the ship's huge form, gleaming dimly under the stars of space. It had looked like the back of a great silver whale.

For a minute or two he had stared absently through the little circular window, and then, hearing footsteps down the corridor, he had turned to see two figures some hundred feet distant moving away from him. They had obviously entered from a side passage and had probably not seen him. One had been this very Hekalu Selba; Austin was sure of it. Beside him had moved a shadow. The Earthman had not seen it clearly, for the illuminating globes burning here during the sleep period were dim and far between.

He had but a vague fleeting impression of a huge knotty form, bent and grotesque. Its arms were so long that its big hands almost dragged on the floor. Its head was very large and bulbous. The pair had seemed to carry something heavy between them, but Austin had not seen what it was. In a moment the Martian had opened a door in the side of the passage and the two had vanished into it.

When Austin had returned to his stateroom, he was not quite sure he had really seen the monstrous horror. Surely nothing like it was known to exist within the orbit of Jupiter! Shelby had thought of reporting the incident to the commander of the vessel, but he had dismissed the idea as too pointless. Now, however, the memory of that vague black form was haunting him. He knew that it was the key, in part at least, to his feeling toward Hekalu Selba.

The Martian had cast his magazine aside. He was patting the soft cushions of the divan on which he was lounging. "Sit here, my friends," he said in his smooth, precise English. "We shall talk, and then perhaps we shall have a little refreshment." The two complied.

"It will be only for a moment," said the girl. "The ship lands in an hour, and I haven't gathered my things together yet."

Shelby was intensely interested in this queer individual, about whose personality there lingered a strangely indefinable web of mystery—of evil, almost.

"So you too have a passion for mechanics," he said. "Somewhere I am sure I have heard of you before. Kelang Aggar, an instructor of mine at Taboor, spoke occasionally of a young Martian student—"

"Kelang Aggar is my friend," Hekki broke in. "He assisted me with several experiments. But they were nothing—a new alloy, very hard, and having a high point of fusion. The heads of the Space Ship Construction Company said it was ideal for rocket nozzles, but they paid me a mere pittance for the invention. This, and a few even lesser ones are my sole accomplishments in the line of mechanics." Hekalu Selba laughed lightly.

"Let us talk of other things, my friends," he continued. "Let us allow our minds to ramble. See those two beautiful potted palms over there—children of the deserts of Earth, and beside them the slender graceful stem of the purple Kelan, dug from the marshes along the Selgur waterway of my own planet. I have seen them both in their native habitat, waving their fronds as though in cadence with some great silent symphony of the universe. See that tapestry over yonder, with the beast woven into it?"

Here Janice Darell pointed up toward the flattened glass dome that roofed the room. "There is old Mother Earth looking down at us, and the sun is peeping around her rim," she said. "See how the light of Sol sifts through the terrestrial atmosphere. There is a streak of red, of gold, of opal, and beyond are the stars and the blackness of space."

"The contrast of the forces of darkness with those of light," Shelby put in softly.

Hekki was smiling absently. "There are many contrasts," he mused. "The contrast of life and death, of power and weakness, of nightmare and reality."

Words popped into Austin Shelby's head, and, carelessly, he uttered them without thinking: "You often walk with your nightmares, don't you, Hekalu?"

A hard light came into the Martian's eyes as he stared straight at the Earthman. "Perhaps," he said, "and you, Mr. Shelby, often walk in your sleep!" But apparently the incident was immediately forgotten.

Austin wondered how much the girl knew about the luxurious Hekalu. A quick glance of intelligence passed between them.

"I'll have to pack now," Janice said. "Won't you boys walk along with me a little ways?" She took Austin's arm as they arose. Hekki fell in beside them. At the entrance to the corridor which led to the girl's stateroom they paused.

"My business will occupy me tonight and tomorrow," said Hekalu, "so I shall bid you goodbye until, let us say, the following evening, Jan, but if you like I shall have Alka take you home."

"Mr. Shelby has asked permission to perform that little service, you industrious old business man," she replied mischievously. And again that dark shadow flickered momentarily on the Martian's features.

"But you will let me see you the day after tomorrow?" he asked. "I have found a little paradise out at Oak Park patterned after the fairy palaces of my own planet, and besides, I have a new jewel to show you."

"Fie on your jewels, Hekki," she smiled, adopting the stiff trite speech the Martians often fell into. "But anyway, perhaps I shall favor you with my incomparable company. The time you mention is still a ways away. *Sidi yadi*, my friend. Remember I shall be expecting a view-phone call from you soon." Then turning to Shelby: "I'll meet you in the lounge right after the boat lands. Don't fail me!"

"You needn't worry about that, Jan," he assured her.

In a moment she was hurrying up the corridor in the pink glow of the lights. As Austin gazed after her, he could not help but think how wonderful was this fluffy little wisp of blonde beauty. Was she for him? Over her he felt there lurked a dark shadow, but this only strengthened the spell she had cast over him, for it gave to him the pleasure which virile males experience when they know that their loved one requires protection.

Hekki cleared his throat to attract the attention of his companion. When Shelby turned toward him he was fumbling in one of the voluminous sleeve pockets of his blouse. Presently he drew forth a very thin rectangle of a substance resembling ivory, and handed it to the Earthman. Shelby glanced at it. It was one of the name cards commonly used by Martian men. It bore the legend in the interplanetary symbols:

Hekalu Selba, Akar

414 Teldasa

Taboor, P. 4.

Beneath in small letters appeared Hekki's Chicago address.

"I shall want to see you again soon, my friend," said the Martian cordially. "There are many things at my establishment which I would like to show you — much that we can talk about."

Austin Shelby accepted the card and handed Hekki his own. Here was an opportunity to get some first hand information on the mysterious man of Mars and his more mysterious, perhaps sinister doings. The idea that he might be placing himself in a dangerous position, Shelby gave scarcely a moment's thought, for he had in him the spirit of the adventurer.

"Thank you, Akar Hekalu. I shall get in touch with you. And in the meanwhile you can reach me at my address through the view-phone at almost any time for I shall be working on a new mechanism there. *Sidi yadi.*"

"*Sidi yadi*, my friend."

The two men parted.

Fifteen minutes later a rustling whisper was audible throughout the *Ekova*, above the steady purr of the forward-pointing decelerating rockets. It became a deep-toned soughing which rapidly increased in volume to a loud roar, and then to a screeching hiss. The ship swayed and rocked a little. It was tearing its way into the terrestrial atmosphere.

In the conning tower forward, the pilot and his assistant were working calmly and cooly over the bewildering array of controlling mechanisms. Getting those thousands of tons of metal safely lowered into a space ship's cradle on the landing stage, was a difficult task, but the experience and efficiency of the two men was quite competent to cope with it.

Far below was a vast sea of winking lights—Chicago, its colossal skyscrapers looming up severe and white and beautiful in the glow.

The pilot's nimble fingers turned a small horizontal wheel at his side. The liner dipped and dropped slowly earthward toward an area of white light. A mass of cloud poured over the huge hull for an instant and then passed by. The outer shell of the great silvery whale which had been chilled to a degree from absolute zero, by the cold of space had been warmed but slightly by the rapid passage through the atmosphere and now gleamed with jewel-like hoar frost.

Down, down it floated until it was only three hundred feet above the landing stage. A red signal light gleamed suddenly on a panel within the control room, and the wizard of that eerie chamber shifted a tiny lever. The space ship halted and hung motionless supported by its repulsion plates. On the ground in the glare of floodlights white-clad men hurried about. Four mighty arms of metal groped upward from a mass of heavy framework. They clutched the craft with a grating noise, and then, with the slow deliberation of a sleepy giant, they drew it gently down into its cradle.

Within the *Ekova* all was abustle. Its doors, built solidly like the breeches of big cannons, swung open, permitting the cool night air to enter the ship, which for seven days had been a world sufficient unto itself. Gangplanks were let down, and the passengers, jesting gaily with one another began their leisurely descent to the ground. Customs officials worked feverishly. A webby derrick arm pointing out from an opening in the side of the liner, was unloading mail and costly material and equipment sent to Earth from the Red Planet.

The routine processes of debarkation over, Shelby and Janice Darell entered the covered causeway which led to the great terminal building of the Space Travel Company.

The two had caught but a fleeting glimpse of Hekki. He was talking earnestly to a white-clad official, and had not seen them; nor had they tried to attract his attention. Conspicuous among the Martian's numerous possessions was a large basket of metal wickerwork, such as were commonly used to convey dogs and similiar pets from place to place. The sight of that basket had aroused again in Shelby's mind that peculiar sense of the presence of something sinister. Was the monster he had seen in Hekalu Selba's company hidden within that case of woven wire?

Within the causeway was a moving walk which carried Shelby and his companion to the depot. Here the intermittent whirring of pneumatic tube-cars operating in a vast network throughout the city was audible. The young Earthian pair, and the two attendants bearing their light luggage entered an elevator, which carried them swiftly to the landing platform for atmospheric craft on the roof of the building.

Shelby presented his identification tag and gave the number of his plane to the official in charge. The man led the way to a hangar at the side of the platform. Shelby had sent an order by radio to the Sutherland Aircraft Company a few hours before, and, complying with his request, a bright new flier had been delivered and housed here, awaiting his arrival.

The official closed a switch on the wall of the building, and the hangar door rolled open. While the two Earthians were entering the craft the attendants quickly placed the luggage into the load compartment.

Shelby fumbled with the destination mechanism and pressed the starting lever. The propellers, whirled at high speed by the soundless atomic motor, thrummed softly. In a moment, the plane, unguided by human hands, hoisted itself almost vertically into the night and was off. Unerringly it would carry its occupants to their destination.

CHAPTER II
A STRANGE STORY

Shelby looked down at his companion. For a time she had been strangely quiet. Could it be that there was just a hint of a troubled look on her beautiful face? The young engineer felt himself drawn to her more than ever. He wanted to know more about his new Martian acquaintance, but he disliked to ask a direct question concerning him, for he feared vaguely that it might give her offense.

"Jan," he said, "you look worried. Is anything wrong?"

She shook her head, slowly, absently, without looking at him. "No, I was just thinking." She paused, and then in the same absent manner she continued: "Only Hekalu Selba is back, and I thought I was rid of him."

Reassured somewhat by her words, but still taking care to conceal any hint of the menace he had sensed about the Martian, Shelby asked: "What possible difference can his presence in Chicago mean to you? He seemed to me to be a very ordinary Martian nobleman—evidently supplied with plenty of money, and having no other motive in life than to enjoy himself, and perhaps to help others enjoy themselves. A perfectly harmless individual."

Janice's face grew serious. "You say those things because you do not know Hekki," she said. "Shall I tell you about him? It would relieve me to share my knowledge with someone."

The young man nodded but made no comment.

"Two years ago," she began, "I went to Taboor on Mars to study sculpture. Not long after my arrival at school, in the company of a number of other art students, I attended a ball given at a glorious old palace in the heart of the ancient Martian quarter. Our gracious host was Hekalu Selba himself. I met him, danced with him, and talked with him. From the first he was attracted to me and I to him, and so we were often together.

"Though some of his peculiar affectations were obnoxious to me, I thought that his good qualities far overbalanced his failings. He seemed always kind and considerate in his dealings with all about him; he was well informed on almost every possible subject; he painted pictures and played various musical instruments with a skill that was little short of genius, and his tales of his travels and adventures in the little-known region beyond the orbits of the minor planets could not fail to delight any listeners. Dreamer and brilliant artist—that was Hekki as I saw him then. Effeminate—yes, but brave and resourceful too.

"Our intimacy grew. He made frequent proposals of marriage to me, but I put him off, saying that I was not sure I loved him. I informed Father back here in Chicago of our friendship. His next letter showed plainly his enthusiasm over the idea of the possible marriage of his daughter with this young noble of the ancient Martian house of Selba. 'Get him, Jan,' he wrote. 'He'd be the catch of a lifetime. Why, his total assets would make the treasure of Crœsus look like a little piece of twisted copper wire.' Poor practical old Dad! For once his business judgment was in the wrong. It was well that I did not follow his advice."

At this point Jan's story was interrupted by the sudden dropping of the plane. They had reached their destination. The craft descended vertically and landed with a light impact in the center of a small private roof garden at the summit of a great apartment building.

"Dad won't be home now," said Jan. "He was delayed in New York, and will not appear until tomorrow. There isn't anyone else around here except old Rufus, so we needn't go down stairs. Let's sit over there instead." She pointed toward a quaintly wrought bench beside a splashing fountain. The moon was shining, and the solitary cypress tree cast a spear-like shadow over the pool. There was a faint fragrance of flowers in the night.

Janice and Shelby seated themselves and the girl continued:

"Shortly after my meeting with Hekalu Selba rumors began to come to me. Men died mysteriously, and there were people who made vague hints that my noble friend was responsible. An uncle of Hekki's had made him the principal heir to his fortune—shortly afterward the uncle contracted a virulent disease and passed away. On both planets men that were obnoxious to Hekki were murdered—capable business rivals and people who perhaps 'knew too much.' Always the circumstances of their deaths were peculiar. Frequently they were found in locked rooms to which an assassin could scarcely have gained entrance without breaking his way. But such violent methods had not been used. Never was there a shred of evidence to implicate the noble.

"But I was beginning to see Hekalu's true color. The lavish display of his wealth—his estates and his art treasures, and the endless round of good times he sought to provide, were merely an attempt to cover up his wickedness. One afternoon that I was with him, he was under the influence of the Elar drug. His face was red and his eyes gleamed with a wicked light. He proposed to me again, and when I made an angry refusal he threatened me — said that if there was another whom I loved he would destroy him and me too.

"That, I assured myself, was the end. Hekki tried to make up, but when he found that I would have nothing to do with him he vanished. I think he went off into the outer regions of the solar system again. He was gone for a long time, and I devoted myself entirely to my studies.

"Then suddenly, out of the blue, I received a letter from Hekki. It came from a small village far to the west of Taboor. A gift accompanied it. Hekki informed me that in a valley far out in the unexplored Taraal desert he had run across a ruined city built by the Melbar kings some seventy-five thousand years ago. He hoped to make an enormous fortune from the art treasures he had found there.

"The gift and the small photograph he sent me, I shall show you at the first opportunity. They are packed away now. The former is a dagger with a flexible blade of a shiny black substance unknown to me. It does not seem to be metal. The hilt is a lump of platinum. It is carved to represent some strange animal with scores of coiling tentacles. Hekki says that the object is one of his treasures, found on the site of the ancient city. But I have doubted this. I know something of the art of the Melbar kings, and certainly the dagger does not resemble the products of their craftsmen. The same is true of those wares of Hekki's which my friends have bought. They are strange — belonging neither to Earth nor Mars.

"The picture too is equally puzzling. It depicts a night scene in a desert valley. Jagged hills in the distance and the nearer moon of Mars in the sky. The floor of the valley is in shadow and things there are indistinct. There are shapes there — vast shapes, odd and grotesque. And there is something in the foreground which might be almost human!

"In his letter Hekki asked if he might see me again, and I immediately wrote and told him that I would. To you, Austin, this probably seems a crazy thing to do, but like most everyone who is young, I had a genuine love for intrigue and mystery, even though they might be dangerous things to meddle with.

"Hekalu came to Taboor, but I saw comparatively little of him. He seemed always to be tremendously busy. Sometimes he would be extravagantly jubilant, as though he had met with some tremendous success, or again he would apparently be worried almost to the point of madness. What these emotional changes meant, he would never tell me.

"Several times old Alka, his favorite slave, spoke to me. 'The Master is not as he used to be, Miss Darell,' he would say. 'He works feverishly with odd mechanisms, and every night when he is at home he stares out into space toward the farther planets with his new super-telescope. Always, what he sees makes his face turn white and hard; sometimes, he smiles and sometimes his features look like a devil's mask.'

"And still Hekki's weird treasures continued, and still continue to come from the Taraal.

"A group of men was sent by the heads of the Place of Knowledge out into the desert to investigate. They disappeared. The officials of the Planetary Patrol made only a hasty and unsuccessful investigation.

"On the day of my departure from Mars, after having finished my course, I saw Hekki, believing that it was for the last time. He said he was going back into the Taraal. And then he popped up on the liner. And that, Austin, is all I know about Hekalu Selba. What do you make of it? What is he trying to do out there in the desert?" She placed her hand lightly on Shelby's arm and looked up appealingly into his face. "Can't you offer some suggestions, Austin? You know that when suspicious events are troubling you, a plausible explanation eases your mind even though you cannot know the truth. And I am afraid, afraid that he is deliberately following me to Earth!"

While Jan had been telling of her acquaintance with the Martian, Austin had been staring at a very large Sadu moth which hovered, and leisurely moved about on thrumming gorgeous wings, which spanned fully eighteen inches. It moved from blossom to blossom in a nearby flower bed, delicately sipping nectar. Always its great luminous eyes, which glowed like coals of gleaming fire, were turned toward the pair. Shelby had scarcely noticed it, for he was absorbed with the girl's account; but now, when it edged closer towards them, and then made a sudden mischievous swoop not six inches above their heads, its presence could no longer be ignored. The girl gave an exclamation of revulsion and shrank involuntarily toward her companion. He leaped to his feet, and picking up a pebble from beside the fountain, hurled it at the night prowler.

"You dirty eavesdropper!" he shouted angrily. "The man who brought your kind from Mars for ornamental purposes must have been crazy!"

The moth buzzed up into the cypress tree and squatted there, silently, apparently resting. Only its eyes continued to glare fixedly, almost malignantly at the occupants of the garden. But they quickly forgot about its presence.

"I don't know whether I can offer a sensible explanation for Hekalu's actions or not, Jan," Shelby said. "However, as far as his activities in the Taraal are concerned, it seems quite possible that he did discover ruins there, and is trying to keep other fortune seekers away. The ruins may of course not really belong to the Melbar dynasty. They might have been built by some contemporary race. Just what he is doing among the minor planets, we can't any more than guess at. Probably he's just adventuring like a few other people. And as for his following you to Earth—well, I admit that you *do* seem to be popular!"

"You're making it sound awfully simple, Austin," said Jan. She paused and thought for a moment, and then, with seeming irrelevance she continued: "Haven't you heard of queer clusters of luminous specks recently seen by astronomers not far beyond Mars? They called them meteor clusters, but they drifted about here and there, not following definite paths as meteors should do."

"You're trying to suggest that they are space ships, aren't you, Jan?"

She nodded.

"But they aren't," Shelby assured her. "They don't polarize the reflected light of the sun as space ships do. Besides, where could they have been built? Certainly not among the planetoids. And any place on the planets, the Taraal desert for instance, would be an almost equally impossible site for their construction.

"Think of the enormous crews of men and the vast supplies of food and water and materials that would have to be taken out there into the wilderness. Undoubtedly Hekalu could back such a project financially, but he would be discovered before he had made a fair start, and the Martian Planet Patrol would wipe him out of existence. Still, though I don't think that the luminous specks are man-built vessels, I am equally certain that they aren't meteors either."

"Then what are they?"

The young man smiled and shrugged. "I don't know," he said. The intuitive feeling that unknown, and not too beneficient forces were at work in the ether about, was troubling him again, making his scalp muscles tingle.

For a moment Shelby stared at the ground. "Jan," he said, "I didn't tell you what I saw on the liner. I didn't tell anyone because I don't want to be called a lunatic. But I guess it's all right to let you in on this now. Briefly, during the sleep period, I came upon Hekalu Selba prowling in a passageway aboard the *Ekova*, in the company of a vague thing that may have been similar to that shape in the photograph—long arms, big head, squat and muscular. If we knew what that thing was, and where it came from, the snarl might be half untangled."

Janice Darell's face took on a sudden surprised look. "You actually saw what you say you saw?" she cried. When her companion nodded, she continued excitedly with wide-open eyes. "I still believe that Hekalu knows something about the meteor cluster. And the beast figures in somewhere too. Austin," she cried, "what if Hekki is trying something really great? I know you don't take stock in any such idea, but just supposing he is—what if—"

"Let him try!" the young man cut in. "I almost wish he would! I'm afraid he would get the surprise of his life." He was staring straight at the unwinking, malignant eyes of the Sadu moth.

"What do you mean?"

Shelby drew a small black case from his sleeve pocket and opened it. He took from it a device which looked like a tiny pistol. There were several other odds and ends of mechanisms in the case. "For a year I have been working on a new weapon," he said. "All the parts are completed, and tonight I shall finish assembling them. This little gun is the projector for a new ray which I have discovered—an etheric vibration of extremely short wavelength. A portion of the atomic energy in any solid or liquid substance the ray touches is instantly released.

"You doubt whether it is effective? Well, I can't give you any proof now; I can only say that when I was back on Mars, fooling with my first cumbersome projector, which produced only the weakest of vibrations, I blasted a big hole in the wall of my apartment, and nearly killed the Martian physician who lived in the rooms next to mine. I had a devil of a time explaining the explosion, and narrowly missed getting myself into serious trouble. In a few days I shall try to sell the weapon to the Earth Government. If they are convinced of its value, and I don't see how they can help but be convinced, our friend from the Red Planet will have to be very careful if he tries anything."

Shelby glanced at his wrist watch. "Eleven thirty—my bed time," he said with mock seriousness. "But Jan, there's one favor I want to ask you before I go. Try not to see anymore of Hekalu Selba, Akar."

Janice Darell made a valiant attempt to act the part of one whose pride and sense of freedom had been deeply outraged. "Mr. Shelby," she said, "what right have you to tell me what I shall or shall not do?" But a light laugh broke from her lips and spoiled her bluff.

"There are two reasons," replied her companion seriously. "First, because we both believe that Hekalu Selba is dangerous; second—because

I love you." He leaned closer toward her with the light of eagerness in his eyes. "Oh, I know I'm crude, Jan," he said passionately. "I'm just a clumsy engineer, not a poet or ladies' man. What I'm trying to say to you must seem awfully trite, but anyway, I want you with me always."

"You mean — ?"

He nodded.

"All right, Austin," she said quietly, looking straight into his eyes.

His arms crept around her, and now he drew her gently to him.

Some moments later, in the nearby pergola, the door which led to the rooms below opened, and an ancient negro clad in gaudy pajamas and bathrobe peered out into the garden. He saw the pair and recognized the girl. A happy grin came over his wrinkled black face. "Well, if dat ain't a mos' pretty sight to look at," he muttered. "My baby done come back at las', and dat sho' am a han'some boy she got dar!" He turned, and leaving the door open and the light burning on the stair, descended. Very softly and wistfully he was crooning an old darky love song.

It was an hour later before Shelby's craft whirred up into the moon-bathed night over the winking lights of the city. And at the same time the big-eyed Sadu moth which had been crouching in the cypress tree, rose on its velvety wings and sped away, as though some urgent mission had suddenly claimed its attention.

CHAPTER III
HEKKI'S PROPOSAL

When Shelby reached his apartment, he immediately donned his laboratory smock and set to work. But he had scarcely finished mounting a tiny coil of wire within the hand-grip of his weapon, when the view-phone bell rang insistently.

The inventor pulled off his smock and threw it over the materials on his work bench, so that the person at the other end of the view-phone connection, whoever it was, would not be able to see them. Then he snapped the television and audio switches. The mists in the view-plate cleared, and there before him, as real as though he were actually in the room, sat Hekalu Selba. The Martian's eyes gleamed with suppressed excitement.

"Mr. Shelby," he was saying, "it may seem strange that I should be calling you so soon, but I have something simply colossal to talk over with you. You must come up to my place immediately! I realize that you may be very busy, but this *is* important!" And he added, "It's nothing to discuss over the view-phone. Will you come? — please!"

Shelby was about to make a cold reply, but he checked himself. An intense curiosity gripped him.

"All right, Akar Hekalu," he said. "I'll be there." The switches clicked.

Hastily Austin changed to his street clothes, and then gathered together the material for his weapon and placed them in the wall safe. Only one thing he selected from the jumble of apparatus — a tiny pinkish crystal, without which it was impossible to produce the Atomic Ray. This he secreted in a hollow button on his sleeve.

For a long moment he stared at his automatic, which lay on his work bench. "Better take you along," he muttered at length, " — may need you."

A wizened black-clad man whom Shelby surmised was the slave Alka, met him at the entrance on the landing platform of a quaint Martian tower atop a huge apartment building, and ushered him into an elevator. He was whisked rapidly downward, and emerged into the central light-well which pierced the structure from top to bottom. The barbaric tapestries upon the walls of this tall cylindrical chamber, the tiling of the floor, which consisted of squares and circles and spear points of various colored stone, fitted artfully together, giving an effect of pleasant disorder. And most of all, the smell of strange incense in the air, told Shelby that he had dropped into a little bit of old Pagar or Mars. Evidently the Prince of Selba was master of the entire tower, which, in itself, was by no means small.

Alka led the way down a short passage, and admitted the Earthman to a large sumptuously furnished room, one end of which was softly illuminated by a quaintly beautiful floor lamp. The farther end of the room was in complete darkness. The Pagarian architects had made it imitate the interior of a natural cavern, for where the light approached the gloom, two glassy stalactites gleamed with a scintillant elfin light.

Shelby had but a moment to take note of his surroundings — the dark hangings woven with silver threads, the embossed shield and spear of an ancient Martian warrior mounted on the wall — before Hekalu entered. The young man saw at once that the noble had lost his air of bored languor which he had noticed about him at the time of their first meeting. His eyes flashed with excitement and his movements were quick and cat-like.

"I see that you have come quickly, Mr. Shelby," said the Martian, "and I am glad. Won't you sit down?"

With scarcely a pause he continued: "I have great wealth, my friend, and while your means do not seem to be small, I believe that it would be very convenient to you to have them supplemented. Suppose I gave you say, ten times as many jewels as are in the tray over on that stand?" Shelby looked in the direction the Martian indicated. He saw a flat shallow container of considerable size. At its center squatted a repulsive thing about eight inches high, carved from a clear crystalline substance from which there flashed countless points of icy, wicked fire — a huge diamond!

Heaped around it were hundreds of magnificent red *tabalti*, most prized of all gems. An expert appraiser had recently told Shelby that in two worlds only thirteen of them were known to exist. And now he was being offered all these stones by one who hinted that he was willing to give him ten times as many — an utterly staggering fortune!

Hekalu's words fairly dumbfounded Shelby, but they grated upon his sense of pride as well. Nevertheless, his face gave no hint of what passed through his mind. An angry reply, he decided, was out of place.

"Naturally, Akar Hekalu, you want something in return for your amazing generosity," he said coolly. "Of course, I could not accept your offer under any other circumstances."

The Martian nodded. "I have it from a reliable source, Mr. Shelby, that you are the inventor of a terrible weapon—an atomic ray which might be dangerous in the hands of unworthy persons. Turn the weapon over to me as well as all information concerning its operation and construction, and promise to say not a word more about the weapon to anyone, and I will give you the jewels at once."

A flash of surprise passed across Shelby's face but he quickly masked it. So this was it! But how was it that the noble had learned of his invention? Could it be that Janice Darell was playing a double hand?—his Jan. He dismissed the idea as preposterous and utterly disloyal.

The Earthman rose to his feet and addressed the Martian coldly. "If I have such a device I believe that I can place it in better hands than yours."

Hekalu Selba's face gave no hint of anger; in fact he seemed at the point of laughing. "You have done as I expected you would. Your refusal shows me how patriotic you are and gratifies me very much, Mr. Shelby," he said blandly. "You are as a man of Earth should be. However, there is another side to the question. I have certain plans and to have you at large might endanger their fulfillment. Therefore I must ask you to accompany me on a little trip. That weapon of yours will be well taken care of. Now, kindly raise your hands high above your head." The Martian was pointing a bejeweled automatic straight at the chest of his visitor. "You are being covered from two other points in this room so try not to cause any misunderstanding," he added.

Shelby saw the wisdom of obeying the order for he felt quite certain that Hekalu Selba and his minions would not hesitate to shoot him down. What a colossal idiot he had been! He had sensed a trap when the noble had called him over the view-phone and yet he had taken no sensible precautions!

Hekki was searching him now. His long fingers were moving deftly from pocket to pocket. They closed upon his automatic and drew it forth. "Ah," the Martian breathed, "it's as I thought. You have brought a souvenir. A most worthy precaution. And, now that you are no longer in a position to cause any trouble," he continued sneeringly, "I may as well tell you about my ambition—Oh, it is simple enough; men have thought of it before but none had the nerve or ability to put it over. Briefly it is this—to become Master of both Earth and Mars! My friends are waiting for me out there beyond the Red Planet—waiting for their commander. And there is another little hope—there is a certain beautiful flower of your race—" Here he stopped to allow his captive to imagine the rest.

A hard light came into Austin Shelby's eyes. It was the only outward indication of the sudden tornado of emotions and thoughts that swirled in his mind. This man sought to enforce his will upon the planets! The question of whether he was capable of realizing this tremendous dream or not, the Earthman did not pause to debate.

Fifty years before, Saranov had attempted it, and as a result a score of great cities became shambles. Certainly the present foe of mankind was more powerful than Saranov. The monstrous associate of Hekalu and the flitting specks of light far beyond Mars seemed to bear out the nobleman's boast. And if he somehow got possession of the Atomic Ray! And Jan—What was he going to do to Jan! Certainly it was she to whom he had referred! It was this last idea which hammered on Shelby's brain hardest of all. A little fiend within him seemed to shriek. "Escape! Send your weapon to the War Office! Kill Selba if you can, for everything is at stake!" Escape, yes, but how?

"Place your wrists together behind your back now," Hekalu was saying. "I have a pair of magnificent manacles—careful. Do not make an abrupt movement."

A crazy idea had come into the Earthman's mind. He did not expect his plan to work but it was all he could do. With an air of one resigned to his fate, he obeyed the order. He felt the Martian fumbling with the manacles. He was evidently using only one hand. The other presumably still held the automatic leveled at Shelby's back. But it was useless to think of such things.

A slim finger touched the young engineer's wrist. He caught it, twisted it back at the same time, then, summoning all the quickness and force he could muster, he ducked low and hurled himself backward straight into the Martian. There was a loud report. A hot pain seared into the fleshy folds beneath Austin's left shoulder blade. Those hidden in the darkness at the farther end of the room did not dare to fire for fear of injuring their master. Now Shelby was grappling with Hekalu. He gripped the hand that held the automatic.

Two more reports—ineffective, and then the two fell clawing and in a heap on the floor. The shaded lamp was upset and its illumination globes were broken. There was darkness. Shelby heard the shuffle of running feet coming across the marble pavement of the chamber. Help for Hekalu! He'd have to hurry. But the Martian noble, racially much frailer than the people of Earth, was no match for the athletic Shelby. In a moment he was pinned, unable to move. The Earthman tore his weapon from him and thrust its muzzle against his recent opponent's chest. Before he fired he saw the Martian's bold smile; whatever failings Hekalu Selba had, cowardice was not among them.

On the heels of the gun's report Shelby darted from the room and down the short hallway which led back to the central light-well of the Selba establishment. If he could only somehow reach his plane! He gripped the doorknob and shoved fiercely, but the stout metal panels were immovable. He might have known that the outer door would be locked! Oh, what an unutterable ass he had been!

Now what? A hoarse cry of triumph caused him to turn. Alka was racing toward him with leveled pistol. A spray of projectiles spread toward Shelby but the slave's aim was bad and none of them took effect. A split second later Alka pitched to the floor with a bullet through his brain.

But there was another to be reckoned with — one who waddled along rapidly on short powerful legs. Its arms were long and black and more powerfully muscled than a gorilla's. One hand brandished a metal knob-stick, and the other, a long-barreled pistol of Martian design. Silvery armor set with jewels that glittered wickedly in the dim light of the hallway crossed the creature's breast. Its head was bulbous, and its face, set deep in plates of shining black chitin-like armor, consisted only of two enormous eyes and a lipless mouth. No nose at all! The horror Shelby had seen on the liner!

The Earthman fired at the monster. The first bullet clinked harmlessly on his opponent's breast-plate. The second thudded full force upon its skull, but apparently the hard smooth skin of the creature was too tough to allow projectiles hurled from a pistol to penetrate it for it did no real damage — only infuriated the monster. Black hard lids dropped protectingly over its eyes, and its mouth worked convulsively. It quickened its pace and brought its own pistol into play.

Shelby had made a hasty survey of the hall and had noted the stairway beside the door he had tried to open. He darted up this, ducking low behind the stone railing to avoid his weird pursuer's bullets. Perhaps in the chambers above he could find a means of escape. He was leaving a trail of blood on the marble steps, and his wound pained him terribly. He felt sick and weak.

When he had reached the top of the stairs, the unknown horror was already halfway up. It had returned its pistol to its holster. Apparently it had been so maddened by Shelby's shots, that only tearing its quarry to pieces could satisfy its lust for vengeance. And the thing was gaining rapidly!

But the Earthman gritted his teeth and kept doggedly on. He fought back the nauseous giddiness that was creeping upon him. He'd have to escape. Oh God! There was too much at stake — the world and Jan — what was happening to Jan? True, he had killed Selba, but certainly the Martian had minions—men who could carry on without him. He could scarcely have built up all his plans single-handed!

Four flights of steps Shelby and his pursuer ascended. Was there a way of reaching the roof and the plane in this direction? And if there were, could the Earthman reach it before the long arms of the thing so close behind wrapped themselves about him? Such an event, Shelby knew could not mean anything less than failure, and possibly immediate death. The fiend behind did not cry out or order him to halt. In fact it made no vocal sound at all. Not even its breathing, which should have been heavy and labored, was audible. Only the hurried shuffle of its unshod feet. Its silent relentlessness was nerve-wracking.

The engineer saw before him at the top of the stair a small doorway, and beyond it a spiral runway leading upward. The light grillwork gate stood invitingly open. Catching the grill with one hand as he rushed through the door, Shelby sought to slam it shut and latch it. He almost had succeeded, and then a huge hand closed upon the bars. One jerk, and a quick grab with the other immense paw and the strange flight and pursuit would be at an end.

But the jerk was delayed. Shelby fired his last round. It did the monster little harm, even though the distance between the two was but four feet. Nevertheless it caused the armored horror to leap back a step, and the moment thus provided was sufficient.

As Shelby stumbled up the dark spiral he heard the thing below tearing at the closed grill. He knew that it could not delay the thing for long. He had just reached the trapdoor at the top of the long climb, when a muffled ripping crash echoed up dimly from far beneath him. The gate was down!

Feverishly he struggled with the heavy trap. Normally it would not have been difficult for him to lift the rectangle of aluminum alloy; but wounded as he was, forcing his numbing limbs to obey him required almost super-human effort. When he had at last succeeded in hoisting it on its hinges, he could again hear the soft padding of hurrying feet.

The engineer found himself in a large room, one wall of which was curved, conforming to the outer contour of the cylindrical tower. Scattered illumination globes gave a dim light to the place. The room was evidently a storehouse for Hekalu's laboratory supplies. Complex mechanisms stood about, evidently waiting to be installed. There were hundreds of metal drums presumably containing chemicals. There were bolts of heavy fabric and stacks of ingots neatly corded. Set in the ceiling of the chamber were several circular windows through the heavy glass of which bright stars shone. Directly above was the roof, and but a few paces distant, the landing stage!

Escape seemed tantalizingly near, but with sinking heart, Shelby noted that there was no easy means of ascent to the roof. He'd have to try to smash one of those windows. But the monster hurrying up the spiral claimed his immediate attention.

Deeply thankful for the peculiar eccentricities of Martian architecture, he hurriedly proceeded to pile ingots on the closed trapdoor. Each of these ingots weighed well over a hundred and fifty pounds. Fortunately for the wounded Earthman, the distance he had to carry them was only a few feet.

CHAPTER IV
CAPTURE!

He had transferred five to their new position before his pursuer arrived beneath the trap and began to push upward mightily upon it. Shelby transferred several more ingots to the pile just to make sure that the monster could not enter. Then, fighting off the diaphanous veil of unconsciousness that was trying to drop over him, he looked about for something with which to effect his escape.

A long bar of metal caught his eye. He seized it, and with all his strength thrust upward at one of the ceiling windows. But the thick glass, crisscrossed by rods of metal, was not easily shattered.

A rattling noise attracted his attention. He glanced back toward the trap. His pile of ingots was trembling as if shaken by a miniature earthquake. The door was rising upward! It settled back and rose again. An inch crack appeared, and through it Shelby could see two eyes and the muzzle of a pistol. He leaped out of range just in time to avoid the bullet that whizzed across the room and flattened itself against the wall.

He darted around toward the hinged side of the trap, where he knew that the black horror could not fire at him, and devoted his attention to another window. He would have reinforced the barricade with more ingots, but he realized that by spending his nearly exhausted strength that way he would be defeating his own purpose.

A dozen times he jabbed up viciously with the bar before a tiny crack appeared in the round pane of glass. The trapdoor behind him was being shaken violently. An ingot on top of the pile was jarred from its place and crashed to the floor. Yes, the window was giving. A small hole appeared in it.

A pair of shiny black forearms had forced their way from under the edge of the trapdoor. Slowly and mightily the shoulders of the monster surged upward. The door was rising, and this time it did not seem that it would sink back.

Shelby had finished his task. Now, with the upper end of the bar thrust through the opening he had made in the window, and the lower end resting in a slight depression in the floor, he proceeded to climb it to safety. His head and shoulders were through the hole when the monster at last burst its way into the room below. But the thing was just an instant too late to hinder him.

Sweating and bloody, Shelby drew himself to the roof and staggered over to the landing stage. Yes, his plane was there.

The night air, and the flush of success was refreshing him. His exaltation leaped higher and higher as his plane swept him up from the summit of the tower of the mysterious Selba.

A wild refrain was drumming in his mind: "Hekalu Selba is dead! I have killed him!" There was nothing more to do but notify the Municipal Air Patrol—an S. O. S. with his siren would accomplish that. They would raid the tower. If any of the Martian's fellow plotters sought to continue with the project the Earthman's new weapon would take care of them.

Shelby was reaching for the siren button, and then a terrific explosion thundered up from somewhere below, and several hundred yards to his right. He saw the orange flash, and then, in an instant the whole city went dark. Another crash came and another. Shelby saw a dark form glide through the air. From far beneath him he heard a troubled murmur mixed with the din of colliding vehicles. Sirens shrieked. In the distance to his right, a great plume of lurid flame blossomed in the sky.

The low purr of a machine gun sounded behind him, and he heard the almost inaudible tick-tick of poisoned needle-darts piercing the fuselage of his craft.

He zoomed sharply upward for a thousand feet, and then glanced back. There was a dim shadow out there—he was being followed. But this discovery, and the realization that the city was attacked made but a vague impression upon his fast-dimming mind. The warm fluid that oozed from his shoulder, making his clothing sodden and sticky, had all but drained his vital energy.

Somehow he began to doubt that he had killed Selba. It had been only a dream, and the monstrous thing that had sought his life had been a dream too. Hekalu was pursuing him now, trying to kill him! The idea took hold, for he could no longer distinguish fancy from reality. It brought to him a vague fear which would have been completely out of place with him had he not been so near gone from loss of blood. It was like a child's fear of the dark.

He began to fly towards home in a wild zigzag course like a dazed bat, but this favored him, for it enabled him to avoid the darts from the pursuing plane. Luckily he remembered that while under fire combat fliers do not make use of their automatic pilots except as a last resort, for these devices cannot direct the complex movements necessary in dodging enemy bullets. Automatically Shelby watched the guiding instruments and followed their directions.

Several times he signaled with his siren, but no one answered him. Thousands of sirens were hooting, and the Air Patrol was very busy. The darkness, the explosions and the muffled roar from the streets continued.

Two ideas now possessed Shelby's mind and he clung to them with the grim persistence of a wounded tiger. One was to get home, secure his weapon and rush it to the federal authorities. The other was to hurry to Janice Darell.

Presently his plane bounded down awkwardly on the landing platform of the building in which his apartment was located. He stumbled out, and down the dark stair. The elevators were not working. Somehow he found his door and unlocked it. He groped toward the wall safe. It was open, and the little black case which contained the unfinished atomic ray projector was gone. A neat round hole had been drilled in the metal door of the safe.

The view-phone bell was ringing. Shelby stumbled to the instrument and moved its switches. The view-plate did not work but he heard a faint voice which he recognized as Jan's. "Is that you, Austin?" it said. "Can't you help me? Something is out there. It has me cornered in my room. It has killed old Rufus. The house police—" There the connection snapped.

A wild surge of anger quickened the engineer's weakly beating heart. He tried to reach the door, and then he felt a stinging sensation in the back of his neck. A needle-dart charged with a sleep-producing drug had struck him. He slumped to the floor.

A moment later a thing of metal and fabric, fitted with drills and delicate thread-like tentacles, and formed like a giant Sadu moth of Mars, darted out from behind a curtain where it had been hiding. It flew up through the air-tube which had been its means of entrance to the room. On the roof it met a black nightmare, and by means of signs traced in the air with an intelligence that was paradoxically human, it directed the monster to Shelby's apartment below.

The first sensation which bore itself in upon Shelby's consciousness when he was regaining his senses was a terrific throbbing pain in his head. He opened his rheum-plastered eyelids and looked about him. He was lying in a bunk within a small dim-lit compartment. Polished duralumin walls gleamed all about. At the center of his prison was a table, and beyond, built into the opposite wall, was another bunk. There was a black blob of something sprawling on the mattress, but he could not see clearly what it was. The illumination globe in the ceiling was not burning, and only a faint glow filtered through the curtained, circular window. A muffled purring vibration told Shelby that he was aboard a speeding space ship.

Aroused evidently by the stirring of its charge, the thing in the opposite berth arose and strode leisurely toward the Earthian. The metal of its harness tinkled, and sharp points of light flashed against its ebony body, like gems sewn into a sable curtain that is being swayed by a vagrant draft of air.

The Earthman recognized the creature immediately as his recent pursuer. It had pressed the light switch now, and the illumination globe glowed softly. Then the thing bent over Shelby, and with a gentleness that was surprising, it rolled him over and examined his bandaged wound briefly.

The young man conquered his revulsion sufficiently to look up into the monster's face. He thought that it was odd that the sight of it did not terrify him. No, really it was not more hideous than the visages of insects he had seen through a microscope. He studied the hard chitinous visors that blinked over the monster's eyes — the hollow where its nose should have been; and he searched for some hint that there was a human personality within that knotted carcass but found none. The lipless mouth and the blankly staring eyes were without any expression that he could interpret.

Two things struck Shelby as being peculiar — the fact that the monster did not seem to breathe, and the icy coldness of its hands.

The thing walked to the door, unlocked it, and left the room. The engineer heard a grating of the key being turned when the door had been shut.

Taking advantage of the opportunity to move about without being observed, he jumped out of bed and hurried to the window. It was then that he noticed that there was a metal band about his right ankle. A long light chain led from it to an eyelet in the wall. Truly he *was* a prisoner!

A single glance through the porthole confirmed what he had known was true — the black sky and the unwinking stars of space.

There was a narrow walk beneath the window, running the full length of the flier's hull. The railing of woven wire cast a checkered shadow on the walk. Somewhere toward the stern a blazing sun was shining, but Shelby could not see it.

His first thoughts concerned some means of spoiling the plans of Selba's band. He guessed, of course, that they were responsible for his present position, and he realized that it was likely that the zero hour of their attack upon the planets was not far off. Could he escape? — a practical impossibility.

Nevertheless he looked longingly at the emergency space-boat hugging close to the hull of its mother ship, and fitted so admirably into her streamlining. If he could get to the entrance of that boat — it was in some other room farther toward the bow — he could give his captors a run for their money and perhaps reach Earth. And if he did? Shelby had great confidence in the Atomic Ray. He removed the top from the button where he had secreted the pink crystal. It was still there.

But how could he get into the space-boat? Plainly it could not be accomplished now. Perhaps soon — in a few hours maybe, an opportunity would present itself. And there were other things he might do. A moment in the engine room, and he could blow the ship to atoms, and with it, most of the ringleaders of the Selba crowd. Stoically Shelby realized that he too would be destroyed, but if he could serve his world, he would not hesitate to make the move.

Bent on getting as well acquainted with his present environment as he could, the Earthman proceeded to examine minutely everything that was within the range of his senses. He tested the strength of his chain, and began to fumble over each link, without having any definite idea of what value the knowledge gleaned from such a procedure would be to him.

He had reached about the tenth link when he heard a sound above the purr of rocket motors — voices. There were two of them. One was a man's; the other was soft and feminine. Shelby knew it at once — Janice Darell's! So she too was aboard the space flier! He realized it with a pang of apprehension. In vain the Earthman tried to catch the words they were saying, but beyond detecting the chilly tone in the girl's voice, he could get no idea of what they were talking about. Apparently they were in the room next to his.

He heard footsteps in the hall outside, and returned quickly to his bunk. Three people entered the room. The first was the black monster. Shelby gave a gasp when he saw who followed it—Jan. She looked tired and worn but in her face there was no hint of fear. She smiled wanly at Shelby. There was another behind her. It was Hekalu Selba—the man the Earthian thought he had killed! For once Shelby was really dumbfounded. He uttered the Martian's name without thinking.

The noble grinned in Satanic amusement. "It is I, none other, my friend," he said. "Aren't you glad to see me? You look as though you were being visited by a ghost."

The Martian chuckled. "But thanks to a breast armor I still belong to this plane of existence. I admit though that you gave me a great scare when you nearly, but not quite, escaped. My four bombing fliers supplied an adequate diversion for the Municipal Patrol, didn't they? And my Sadu moth, radio controlled automaton—it functioned perfectly!"

Shelby rose from the bunk and sauntered toward his captor. Hekalu made no move to stop him. "Now that you have Miss Darell and me nicely trapped, what do you intend to do?" Shelby inquired coldly.

The Martian laughed. "You have a very inquisitive nature, Mr. Shelby," he said. "What do you expect me to do? Continue with my plans which you so almost successfully spoiled, my friend." Here Hekki's voice became suddenly excited and husky; his lips curled and his eyes took on the fanatical look of a megalomaniac who sees within his grasp his dream of power.

"Very soon," he lisped, "we strike. Mars first, then your planet. I shall be great—greater than all the combined rulers of the millenniums gone by, and Janice here, will share my greatness." The slender arm of Selba stole around the waist of the girl beside him. She did not try to draw away. "That last little idea maddens you, doesn't it, Mr. Shelby?" he added with a sneer.

Shelby felt a flush of heat in his cheeks. What happened to Jan that she should permit the noble to be so familiar with her? Had she been dazzled by his wealth and his promises of what stupendous things the future would bring? For a fraction of a second something seemed to let go in the Earthman's mind, and then he saw the fleeting look in the girl's eyes. He checked the impulse that had urged him to send a fist crashing into the face of the smirking noble. Certainly such an act of violence could accomplish no good.

Shelby looked at the black monster. It was standing beside the table, and leaned forward, so that its knuckles rested ape-like upon the floor.

It was gazing narrowly at the Martian, and its mouth opened and closed nervously. There was a faint something in its almost blank face which suggested to the Earthman that the bond of friendship between the Prince of Selba and this weird devil of the void was none too strong.

Hekalu withdrew his arm from about the girl. He nodded toward the bejeweled nightmare. "I had almost forgotten my lieutenant here, Mr. Shelby," he said. "He is the ruler of the empire from which I am recruiting my forces — my chief ally. Since his people do not employ a language of sounds, he has no vocal name; but for the sake of convenience I have christened him Alkebar, which means 'The Unknown.' He was my companion on my recent trip to Earth, for he wanted very much to see what a beautiful place is your world." There was a sinister hint in these last words.

Hekki made a few quick signs to Alkebar with his fingers, and then turned to the girl. "I must ask you two to leave us now, Jan," he said. "Mr. Shelby and I have an important matter to discuss."

Alkebar grasped Janice's arm with a horny paw, and hurried her through the door. But nevertheless Shelby caught a fleeting glimpse of her face as her lips formed, but did not utter, the word — "Wait." Hekki did not see.

The Earthman turned upon the Martian. "I am going to usurp your assumed right to start this little private conversation, Akar Hekalu," he told him. "There is only one thing I have to say. You are a noble, the son of a long line of nobles who righted wrongs and avenged insults on the field of honor. You have wronged me, no you have outraged me. Therefore I challenge you to combat. Choose your weapons. No place will suit me better than this room; no time better than now." But if Austin had expected to nettle Hekalu into a mood for fighting, he was disappointed.

The Martian was smiling mockingly. "Life is sweet," he said, "sweeter to me than it has ever been before. I do not wish to die — not even by your hands. And you — you have certain knowledge and information which is valuable to me. You must live. I was going to talk to you about what you know. That weapon of yours — we are working on a projector. But something is evidently missing — a tiny element."

"What you have learned about the Atomic Ray," Shelby cut in, "you learned through your own efforts. If you can steal the remainder of the necessary information from my brain, you are welcome. Otherwise, I urgently invite you to go to the devil."

Hekki's face assumed a look of infinite though make-believe sadness. It was a trick such as a designing woman might use to attract some desirable male.

"I am sorry to hear you talk so, Mr. Shelby," he said. "But as you suggest, I believe that there are ways of stealing knowledge even from your mind. For instance, in an old vault beneath my palace at Taboor, I once found a sealed vat containing a certain fluid. The Ancient Ones were wise, for when they desired any man to talk, they thrust his arms or his legs, or perchance his whole body into the fluid. Very slowly, and with some discomfort, it ate away the tissue of his nerves. I must leave you now, my friend. Think well, and may the gods that rule the universe guide you on the right course."

He opened the door. Shelby caught a glimpse of a long hall, and at the far end, the bewildering maze of control-room equipment. The panel closed.

CHAPTER V
THE RACE THROUGH SPACE

Immediately the Earthman set himself to the task of examining everything in his prison. But as he had expected, there was little or nothing to discover. The walls which his tether permitted him to reach were all perfectly smooth and solid. He realized with a sheepish grin that it had been foolish of him to even dare to hope that they would be otherwise. The chain fastened to the fetter was quite adequate to hold him. The window, even if it might have been used as an avenue of escape, was securely fastened with bolts, so that it would have taken a man equipped with a heavy set of wrenches, an hour to remove it. To shatter the flexible pane was next to an impossibility. The table was firmly welded to the floor. Beyond the table, Shelby could not go, for the chain prevented him. But he was quite sure that there was nothing movable in the entire room massive enough to be used as a tool or weapon.

He slumped down on his bunk, and let one hand rest on a small power-pipe which ran along the wall and up to the illumination globe above. For a minute dejection almost got a firm grip on him. But he fought it off. This was no time to give up. Why, the struggle hadn't even started yet!

Shelby felt a faint vibration of the power-pipe under his hand. For a considerable time the impressions had been coming to him, but they had scarcely penetrated into his consciousness. They seemed no more significant than the hundred and one little noises and disturbances that go with the running of any space ship. Presently however, the regular sequence of the pulsations attracted his attention. Something made him think of the almost obsolete Morse code. Then the realization came to him. Someone in another room on he ship was tapping on the power-pipe—signaling—signaling him! He spelled the word out—A-u-s-t-i-n, repeated over and over again.

His first thought was of Jan. It must be she who was calling him for there was no one else.

Quickly, with his heavy signet ring, he tapped out an answer: "It is I, Jan, A. S. shoot—"

With tensed muscles, and with fingers firmly clutching the power-pipe that he might not miss a single signal, Shelby crouched, receiving the message. Somehow there was an urgency, an insistence, an appeal about those hurried pulsations that no human voice could have conveyed. It was fantastically like communicating with one who is buried alive.

"We must escape not later than five hours from now," the tapping spelled. "You have been unconscious for a long time—drugged. In five hours we land on Mars. Then escape will be impossible.

"Hekki has told me much, and I have seen much. The horrors that are Selba's henchmen—three times some of them came to the ship, once in a band of over a hundred. Hekki is worried. He has not troubled me yet. Too busy I suppose. I have tried to make believe that I agree to his plans. I thought I could control him that way. But he has been taking the Elar drug.

"We must escape, Austin. We must! Can't you think of a way? I will help! If they get you to the concentration base in the Taraal they will torture you. And we must remember our homeland!"

The hurrying vibrations ceased, and then, almost before he knew what he was doing, Shelby was tapping out an answer promising the impossible.

"Never fear, dearest," he signaled. "Just let me think for a few minutes." A moment later this phrase almost made him laugh. The sap hero of a comedy which had recently been broadcast over the radio-view had said almost these exact words. Think? Of what? Escape within five hours? How? But Jan's appeal sent in such an odd way had an almost magical effect on him, and made his brain work harder almost than ever before. And then the ghost of an idea came. There was a chance that it would work. He signaled to Jan, and then for half an hour, they put their heads together—planning.

Somewhat nervous, Shelby walked to the door and hammered loudly upon it. A thin-faced slave whose hide was burned by desert suns to the color of mahogany, appeared almost immediately.

Shelby answered his inquiring look briefly: "I would speak to your master," he said in Pagari—"right away." The slave nodded and reclosed the door.

In excited impatience the Earthman waited. Now and then he tapped short messages of encouragement to Jan. Would Hekalu never come? The strain of suspense was not exactly pleasant. Finally, unable to contain himself any longer, he rose from the bunk where he had been reclining in readiness for the first move of the coup he was planning, and began to pace the floor.

He chanced to glance out of the window. On the railed walk beyond, a man clad in space armor was bending over a small portable case which was supported on a tripod. Shelby surmised correctly that this man was Hekalu Selba.

Beside him, paying close attention to whatever the Martian was doing, stood the black Alkebar. The Earthman frowned in puzzlement, almost in awe. For Hekki's weird companion wore nothing that would be of the least help in protecting him from interplanetary cold and lack of air pressure. Not even an oxygen helmet! And yet, as the monster examined interestedly, every dial and switch that Hekalu touched, he showed not the slightest hint of discomfort. The airless emptiness of space seemed home to him. How could such things be? A strange thrill tingled and vibrated along Shelby's spine when he realized how alien was Alkebar. There was no kinship between him and the creatures of either Earth or Mars.

Presently Hekki looked up, and as though moved by some intuitive realization that he was being watched, turned awkwardly in his cumbersome attire, and glanced along the row of portholes in the side of the vessel. He saw the Earthman and smiled at him. Shelby felt that it was the kind of smile which a tolerant father might show to his youngest son. Hekalu waved his hand, and his lips, behind the glazed front of the helmet, formed several words which Shelby could not interpret. Then the Martian returned his attention to his apparatus.

When Selba entered his prisoner's room some moments later, he found him lounging on the bunk.

The Martian looked enquiringly at Shelby. "You have reached some conclusion, my friend?" he asked.

Without changing his position on the bunk the young man nodded. There was an expression of dejection and sullen resignation on his face which he was trying hard, above the intense excitement which possessed him, to make realistic. Still acting the part he spoke: "Yes, Akar Hekalu," he said between teeth that were apparently gritted with rage, "I have decided to reveal to you the secret of the Atomic Ray."

A triumphant gleam came into the Martian's eyes. "Ah, my friend," he said, "you at last see the light. I knew that you would. But what has been the cause for this sudden change in attitude? The torture chamber, perhaps?" There was an undercurrent of suspicion in Hekalu's voice.

Shelby turned his head sullenly away, feigning shame. He said nothing. A minute passed during which time Hekalu stared at his captive, a sardonic smirk of contempt curling his thin coral lips.

Finally he said, "I will have Koo Faya bring you writing materials, and you will describe in writing every detail of the manufacture of the missing element."

"No," replied Shelby, turning his face toward the Martian, "I haven't the ability to do that. It will be necessary for you to take me to the laboratory of the ship where I can demonstrate the process to you. It is much too delicate and complicated."

The noble's eyes wavered slightly. "Once," he said, "you tried to trick me, but I warn you that I am on guard now so do not attempt it again."

He signed to Alkebar who had been standing silently beside the open door. The giant drew a key from a pouch at his side, and kneeling, unlocked the fetter fastened about Shelby's ankle. It rattled to the floor. And at the same time the Earthian, leaning back on the bunk with arms stretching over his head, tapped sharply three times with his signet ring on the power-pipe. It seemed to be only an unconscious gesture—nervousness perhaps.

Immediately there was a terrific crash from down the passage way, followed by an agonized scream. Another crash. More screams.

Hekalu started, and then making a hurried gesture to Alkebar which indicated that he was to guard the inventor of the Atomic Ray, he drew his automatic and dashed down the corridor to investigate the disturbance. The Earthman however, was in no mood to be guarded. No longer shackled, he leaped to his feet and over to the center of the room. The great voiceless beast from the stars stood before the doorway with his long arms outstretched. He was not trying to capture the Earthman—only seeking to block his path.

But Shelby had no time to waste. Gathering himself together, he hurtled straight for the ankles of his opponent. The fact that the artificial gravity of the ship was of the same strength as that of Mars—only a trifle more than one-third that of Earth—added to the effectiveness of his plunge. The mighty-muscled Alkebar, puzzled by the unheard-of tactics of his agile though vastly weaker foe, suddenly found himself in a sprawling heap on the floor. Shelby leaped over him through the door, slammed it, and raced precipitately down the corridor.

In the meantime Hekalu Selba had reached Janice Darell's room, but when he had unlocked it and had thrust his head inside to see what the matter was, a heavy urn, deftly aimed, had crashed full into his face. Shelby saw him sprawling in the passage badly dazed, and a split second later Jan dashed from her cabin. She looked around, and when she saw Shelby coming swiftly toward her she flashed him a quick smile of triumph.

But Alkebar had wrenched the portal of the Earthman's recent prison open, and was in hot pursuit. He was tugging frantically at the pistol in his belt.

"Run, Jan, quick!—To the control room!" Austin shouted.

He caught up Hekki's automatic which had dropped from the Martian's grasp when he had fallen, and wheeling, fired at the black colossus. The bullet struck Alkebar's right hand with which he was raising his pistol. The tough natural armor which covered the monster from head to foot prevented it from doing any serious damage, but it must have stung badly, for his weapon clattered to the floor. While he was stooping to recover it, Shelby hurried forward to catch up with Jan. It was but a few yards to the control room. If they could get there, overcome whoever was in charge and barricade themselves in, they could master the ship!

Their luck had been good, but it was not destined to be as good as that. They caught but a brief glimpse of the bewildering array of switches, dials and levers, that constituted the brain-center of the craft. Standing on guard before his instrument panels was the mahogany-colored slave Koo Faya. He was half crouching, at bay. There was a murderous light in his eyes, and he held leveled in his hands a light machine gun. Shelby's automatic was leveled too, and he pressed his trigger an instant before the Martian. Four bullets whizzed into the control room, splattering close about the thin mummy-like body of Koo Faya. A glass globe that glowed redly on the top of a complicated mechanism, was struck and burst with a popping sound. A rose-colored vapor floated ceiling-ward.

Simultaneously Koo Faya's weapon began to whir. Then, even as Shelby jerked Jan back out of danger, the wild shriek of an alarm siren mingled with the discordant clashing jangle of ungoverned machinery running amuck, rang through the ship, and the huge metal cigar pitched and careened like a frightened thing.

Alkebar, having recovered his pistol, was staggering down the passage shooting rapidly. But owing to the crazy motion of the space flier his missiles were momentarily not taking effect.

Austin and Jan knew that Koo Faya was leaping to a position where he could shoot his poisoned darts at them again. What now? Cornered? No! Janice Darell wrenched open a door in the side of the passage and shoved Shelby into the tiny room beyond.

In the opposite wall of the closet was a round dark opening. "The emergency flier," Jan shouted. "Into it!"

As quickly as they could they climbed through into the submarine-like interior beyond. Fighting to keep themselves erect, they slammed the heavy duralumin portal to and fastened it. Alkebar was already groping on the opposite side. But he was too late.

Shelby leaped to the control panel and cut the electric current from the magnets that held the emergency flier anchored to its mother ship. It floated, free from the careening hulk. Its rocket motors roared into life.

The occupants of the tiny craft looked back at the *Selba*. It had ceased its mad motions now, and was hanging quietly in space. Evidently Koo Faya had succeeded in righting matters to some slight extent at least. Would he be able to patch things up entirely? The red globe could be replaced in half an hour. It would be that length of time at least before the *Selba* could engage in pursuit.

But the arm of a space ship, equipped with weapons commonly used in the void, is long. Hence Austin Shelby considered it his first duty to put as much distance between his craft and Hekalu's ship as possible.

Still four million miles away, Mars glowed—a tiny red disc; and he headed toward her giving the flier full freedom to do its best. The fiery vapors fairly tore from the rocket nozzles.

With one hand in readiness on the control lever, which resembled in appearance and operation the joystick of an airplane, and his feet on the bar used for steering in a lateral plane, he kept his eyes fixed on the receding bulk behind. Jan had handed him one of the two pairs of binoculars which she had just found in the supply compartment.

Austin knew what to expect from the direction of the *Selba*, and it came well within schedule. A flash of green fire spurted from the foredeck of the ship. It showed up with startling vividness against the jeweled sable of the void.

Abruptly Shelby drew the control lever back. In response to his movement the rocket nozzles, now deflected from alignment with the central axis of the craft, sent it into a steep climb. The terrific angular acceleration seemed intent on forcing the two fugitives straight through the metal floor. It drew the blood from their faces and made them grow pale and giddy. But they escaped being struck by the torpedo.

It exploded a hundred yards beneath the flier's keel. Fragments of it banged against the hull. In rapid succession other flashes darted from the *Selba*, which had dwindled to a silvery speck far to the rear. But still those missiles, directed by incredibly delicate sighting mechanisms, and hurled at almost the speed of light, continued to score remarkably close to their target.

If it had not been such an elusive target they most certainly would have blasted it to fragments. But Shelby, skilled as were most of the men of his time, in the handling of small space craft, was able to endow his flier with much of the agility of an alarmed dragon fly. Darting, weaving, zigzagging, yet always keeping its general course fixed toward Mars, it careened away. Always it was ringed by an aura of green flashes.

However, good fortune is seldom perfect. The tempered duralumin plates of the flier managed to withstand the force of all of the torpedo fragments which showered them—with one exception. One dart from Hekalu's ship exploded barely fifty feet to the right of the fugitive craft, and a flying chunk of steel sent it pitching and tumbling through the ether.

When the two bruised occupants had regained their equilibrium they heard a faint hissing above the roar of rockets. They knew that there was but slight chance that the *Selba* could do them any further harm, for though the torpedoes continued to come, the distance between the two vessels was now so great that a damaging shot was almost an impossibility. Nevertheless, the present situation was serious enough. A leak!

Fixing the nose of the flier toward the Red Planet, and locking the controls, Shelby left the pilot's seat to determine the extent of the damage, while Jan searched the supply compartment for something with which to repair it. There was a deep dent in one of the ceiling plates and a thin wriggly crack through the center of it—not an easy job to patch that out in space under the best of circumstances.

The young man whistled when he saw how near they had come to a hideous death. Several times he had seen the bodies of men who had been suddenly exposed to the pressureless airless cold of the outer void—hideous bloated things through whose skin the livid blood had forced its way.

"Any luck, Jan," he asked, looking back at his companion. "Did you find some cement?"

She shook her head.

CHAPTER VI
THE SPACE MEN ATTACK

First stepping to the oxygen supply valve and opening it a trifle wider, Shelby hastened to assist the girl in her quest. Their ears were ringing. The air pressure within the hull was dropping rapidly. Diligently they ransacked every nook and corner, but found nothing more valuable than a can of thick grease. Shelby smeared some of it over the crevice; it helped but did not by any means check the flow of the escaping air entirely.

"It's a race with time now, Jan," he said quietly.

She looked at him. Her face was a trifle pale, but her lips and eyes were smiling. "Are we on our way to Mars, Captain?" she enquired.

He nodded. "We are, Admiral. The fuel tanks are full and if our air lasts we'll get there."

"And when we do," she put in, "the best of luck to Hekki and his friends!"

A vision swept through Shelby's mind—batteries of fantastic machines whose maws spewed flames of faint lavender fire—blinding flashes of light and world-rocking explosions: a hideous thing to dream of—hideous yet glorious, for the civilizations and freedom of two worlds depended upon it. To the Red Planet—they *must* make it!

Janice Darell had placed her hand lightly on Shelby's arm. Her expression was serious, almost hard. "Austin," she said, "tell me truthfully, can we really reach Mars? It is likely that we shall get there before we go out?"

"Certainly, darling," he replied, putting as much assurance into the words and expression as was possible. "Why do you ask?"

There was something that suggested doubt, perhaps even displeasure in her answer: "We have a duty to perform, Austin—a duty infinitely bigger than our own petty existences. You have not seen what I have seen—small scouting patrols that came to the *Selba* riding strange round things that must have been machines of some kind. One look at those henchmen of Alkebar, their great black bodies, their quick nervous movements—like eager panthers, their wicked-looking weapons which they carried with such an air of easy assurance, and you would have known what they hoped to do. Most of these devils are within the orbit of Mars for the first time. Certainly Hekki has told you something about them?"

Shelby nodded. "Very little; but I have noticed a few of Alkebar's remarkable peculiarities," he said.

"Well," she continued, "if we can't get to Taboor, there is one thing we can do — destroy the *Selba*, and with it Hekki and Alkebar."

"Destroy the *Selba*!" Shelby exploded, "with what? Those toy machine guns on the nose of this bus? The bullets wouldn't even make noticeable scratches in the hide of that tough old girl."

"Not with the machine guns," Jan said slowly, "with this flier! A little luck and it would work."

The idea flashed through Shelby's brain. Ram the *Selba* at high speed! Absolutely certain self-murder! A wave of tremendous admiration for the girl came over him. She had something more in her favor than mere beauty and intelligence.

"Your idea is a pretty good one, Jan," he told her. "But rest assured that unless you can overpower me, it will never be put into execution. However, I'll tell you the truth: we have about a fifty-fifty chance of reaching the Red Planet alive."

And so they tore on their way across the void while they watched the dial on the oxygen tank. They were racing with a tiny needle that crept ever nearer to the zero point that was its goal.

By allowing the pressure within the flier to drop to the lowest point that they could endure, they managed to conserve considerable oxygen, for then the rate of escape from the crevice the torpedo fragment had made was naturally not so rapid.

Frequently they examined the sky behind them, expecting momentarily to discover the tiny speck of flitting silver that would be the *Selba*. But if the ship was pursuing them it had not yet come close enough to be seen.

However, there was another, and perhaps greater menace which kept their eyes turning this way and that, searching for signs of danger. Clusters of dully-glowing specks in any quarter of the heavens would be the first indications of its presence. They would grow larger, come hurtling on like racing meteors in the sun's glow. Only there would be an odd wobbly motion about their darting flight. Shelby tested the trips of the two machine guns. Spurts of green flame plumed out of the muzzles.

He had set the radio transmitter in operation, and was sending occasional signals for assistance. But he knew that this was practically a useless move. Hekalu had taken them far off the beaten track, and they were still half a million miles from the Terrestro-Martian traffic lane. The range of the transmitter of this craft was only ten thousand miles. Even if they had been much nearer the chances of their signals being picked up were slight.

The Martian disc was growing larger. It had become an ochre sphere delicately ringed and mottled with greens and browns like a cloudy opal. The flier was fairly eating up the distance.

Shelby had just said: "I believe we're going to make it, Jan," and then the signs which they had hoped would not appear came. Ahead of them and a little to their right, a vague cluster of specks glimmered into view. It wavered like a wisp of luminous smoke buffeted by a light breeze. This was the one thing that distinguished it from a meteor cluster.

Rapidly the individual points of light grew, becoming tiny stars that glowed by the reflected light of the sun. Within five minutes there was no longer any chance of mistaking their identity, for their flat disc-like shapes and the half-human forms of the things that rode them were already visible through the binoculars. They were approaching at terrific velocity. Both Jan and Austin knew them to be subjects of Alkebar. There was no mistaking their motive. Doubtless orders had been flashed to them from the disabled *Selba*.

Realizing that these fleet space riders could easily catch up with his flier if they so chose, Shelby made no attempt to elude them. Instead he clung doggedly to the straight course toward Mars.

The twin machine guns, responding obediently to their directing mechanism, swung on their swivel toward the hurtling foes. Shelby peered into the eye-piece of the "sighter," a complicated arrangement of mirrors and lenses which enabled the pilot to always look directly through the ring-sights regardless of what direction the gun barrels were pointing. He pressed the trips, and soundlessly, out in the vacuum of space, the guns went into action. Flickering green flames of detonating radio-active explosive darted from their muzzles.

Almost immediately there were answering flashes among the approaching shapes, for the high-calibre bullets were also loaded with explosive. One projectile took effect—another! Emerald flares of light, and nothing remained of two bold space men and their queer disc-like vehicles but torn fragments of flesh and metal.

The Space Men were very close now. Jan and Shelby could see the light flashing on their jeweled harnesses and on the weapons which they flourished defiantly. There must have been almost five hundred in the party. Somehow their wild charge was vaguely reminiscent of a band of fierce Bedouin marauders, racing madly across the desert, bent on pillage. Only it was the Arabs who suffered by this comparison, for the desert of these mysterious Space Men was the whole of interstellar emptiness; and their forms and those of the things they rode, were the forms of the forces of Iblees himself.

Apparently these henchmen of Alkebar had some object in view other than the mere destruction of the flier, for they made no move to use their weapons. They were pulling upon levers on their vehicles, checking their headlong flight.

Now they were coursing with the little craft, swarming about it, edging nearer, at the same time taking care to keep out of range of Shelby's guns.

There was a scraping against the hull and a light jolt as a talon secured a hold on an eyelet ring. A black bulk dropped down on the nose of the craft. A pair of hands gripped the barrels of the machine guns, and with an easy tug, tore them from their mountings. There were shifting scratching sounds coming through the flier's light shell—heavy bodies moving about, and then a sudden ripping vibration. The control lever felt loose in Shelby's hand. He could no longer guide the vessel. And there was nothing either he or Jan could do except wait. The rocket motors still purred evenly.

"I guess they've got us this time, Jan," the young man said to his companion. "I wonder what they are going to do with us?" He spoke as casually as though this latest unfavorable turn of fortune was no more serious than the loss of a game of chess.

Janice Darell was equally cool. "Next time we win," she laughed. It is odd how human beings so often react to strange and terrifying situations. "I'm always ready, you see. Here I was crouching behind you throughout the fight with this perfectly useless pistol in my hand, hoping foolishly that I might be able to use it. That's loyalty."

They fell to studying the two monsters which rested on the nose of the craft in front of the pilot's observation window, where the guns had been. The Space Man was crouching out there trying to peer in at them. He was very much like Alkebar—only not so large, and his equipment and adornment did not boast so many jewels.

Shelby felt a peculiar sense of the unreality of the creature. He looked into its face and saw its eyes. Beside the left orb was a mottled area that must have been a scar. It seemed as concrete as anything he had ever seen, and yet for the second time, he told himself that such a creature wasn't possible!

Time honored tradition had said: "Life can exist only where there is oxygen, water and warmth." And all three of the requisites were lacking in the void. Shelby realized that tradition might be wrong, but the question still remained: How did these creatures of space live? Whence came the energy that kept their bodies functioning? If not from the combustion of food with oxygen, then where? If there were no moisture in their bodies, and there certainly couldn't be, for it would have been frozen in an instant and diffused through sublimation, how could vital fluids flow through their veins? He put these questions to Jan, but she shook her head.

"Hekki informed me that these people inhabited a region somewhere beyond Mars, but he did not tell how it was that they could live in space," she said. "It might be that they have had a development similar to terrestrial insects with the skeleton of armor enclosing their flesh."

The vehicles of the Space Men were even greater puzzles. How did they fly out here where the rocket was the only human invention that could move? Many of the vehicles were visible now through the flier's windows. They were disc shaped platforms of a strange lusterless metal. In the center of the top was an opening in which the Space Men sat. Projecting from the discs were a series of levers, permitting evidently simple control. But no hint of their principle of operation was given. They emitted no rocket jets; no beams projected from them.

Austin realized that there were many mysteries of the universe with which he was not acquainted; this was certainly one.

The sound of bodies moving about on the outer shell of the flier was still audible. Presently there was a sharp explosion somewhere toward the stern. The rockets immediately fell silent. The fugitives saw that some of the Space Men were now busying themselves with long metal cables. Deftly and expertly they were looping them through the eyelet rings set at frequent intervals along the sides of the flier.

The other ends of the cables they fastened firmly to similar rings on their vehicles. They finished the job with all the efficiency of trained military engineers. Then, with the small interplanetary vessel in tow, the Space Men began to move off toward Mars, rapidly gaining momentum until their speed must have considerably exceeded that which most space craft could equal. They deflected their course somewhat from the direct path to the Red Planet, probably to avoid a meeting with any wandering ship.

Throughout the fantastic voyage Shelby and Janice Darell found little to do but stare dumbfounded at their weird captors and to watch the rapidly dropping needle of their oxygen supply-gauge. But as it proved, there was little danger of suffocation, for the Space Men were making good time.

And so, after two hours of flying they came to Mars — not to Taboor which the fugitives had previously hoped to reach, but to a deep valley in the desert of the Taraal. The strange caravan circled around to the night side of the planet, and then, slowly and carefully, but with a hint that they understood their work well, they proceeded to lower the disabled craft through the atmosphere to the ground below.

The door of the flier was torn open like a paper thing, and a black giant fully as huge and burly as Alkebar himself hustled the adventurers roughly out into the open.

The pock-marked face of Loo, the Martian name for their nearer moon, was in the sky, and by its light they could see hundreds of Space Men crowding about them. Plainly this Martian colony was fairly well peopled, for there were many more than the five hundred who captured them. The attitude of the onlookers was one of casual curiosity. For the moment at least they were not showing the more brutal side of their characters.

The fugitives were given but a moment to look about, while their jailer apparently carried on a silent conversation with one of his lieutenants.

They saw the sandy floor of the huge rectangular enclosure dotted with strange mounds which must have been some kind of shelter, the encircling walls crowned by square towers at regular intervals. Those walls were amber-colored in the moonlight, and cast dense shadows that shifted visibly as Loo raced in its meteoric course toward the east. Here and there before the mounds huge vague shapes squatted. At the center of the enclosure a tall spire of silvery girders rose, supporting at its summit a cone of a dull black substance. It looked like the creation of either Earthmen or Martians.

Beyond the wall the rounded summits of desert hills, over which in ages past, a restless ocean had poured and flowed, were visible. In spite of their position the two young Earthians could not help but marvel at the silent grandeur of this exotic scenery. A light though chilly desert wind blew refreshingly against their faces.

The black giant had kept a hand on each of his prisoners during his brief conference, and now, none too gently, he guided them to the entrance of one of the mound dwellings. The Space Man ushered his charges into a corridor, and then, fumbling with a curious lock he opened a heavy door and shoved them into the dim-lit room beyond. With a rattling clink the great stone panel closed behind them.

A lump of self-luminous rock set in the stone ceiling gave a faint illumination to the bare interior. There was no furniture — only the sand-covered floor and rough rocky walls. On the floor a Space Man, larger and more magnificently-muscled by far than any they had yet seen, sprawled.

He was either unconscious or dead; they could not tell which. There were hideous welts and gashes and half-healed scars all over his body. The gashes were caked with a viscid purplish substance.

With the coming of the sudden Martian dawn which flashed through a narrow embrasure high in the wall, the jailer returned. His first act was to thrust the needle of what appeared to be a form of hypodermic syringe into the arm of the unconscious Space Man. Then he led his Earthian captives out into the open.

Neither Jan nor Austin were surprised when they saw the *Selba* squatting near the base of the spire. Several Space Men, directed by the slave Koo Faya, moved about the ship, working the fueling pump.

Walking down the gangplank which led up to the entrance of the vessel was Alkebar, and beside him, Hekalu himself. The latter sauntered leisurely toward his captives, and the Chieftain moved off toward a group of Space Men standing some distance away.

CHAPTER VII
ANKOVA'S STORY

The Martian made a brief nervous sign to the jailer. "Gently, Rega," he said. The Space Man relaxed his painful grip on his prisoners. The noble surveyed them smiling. Defiantly, half contemptuously, Shelby was smiling back.

Finally, with a mocking casual air, Hekki spoke: "There is a very ancient saying on your planet," he said, "to the effect that bad pennies always return." The corners of his mouth twitched with sardonic amusement. His manner grew more serious, yet still there was an undercurrent of sarcasm: "Miss Darell and Mr. Shelby, I want to compliment you on your remarkable cleverness and daring. Words cannot express my admiration for you. You have every right to be proud of yourselves."

Shelby nodded. "We are," he told him drily. "Is there anything more on your mind?" He turned away with an expression of bored contemptuous indifference.

"I have little to say except that we are about to continue our recently interrupted journey tonight, Mr. Shelby," said the Martian.

He saw the Earthman and the girl casting interested glances at the disc vehicles that surrounded them everywhere.

"You like my people?" Hekki inquired. "You find them entertaining? Perhaps you have discovered things in their habits which you cannot understand. Shall I give you explanations?" For the moment at least there was a serious earnest ring in Hekalu's voice.

"Flag of truce, Jan. This should be interesting," Shelby said. His eyes were full of eagerness as he turned back toward the Martian. "How do they live out there?" he cried. "There isn't any air or water, and it's almost as cold as it can get anywhere. Why, the thing is utterly impossible according to the laws of common sense!"

Immediately all of Hekalu's lazy air of careless mockery was gone, and the dynamic aura of the tireless experimenter and inventor that had hidden beneath it showed out clear. His voice was husky with suppressed excitement when he spoke:

"I too was dumbfounded when, some five Earth years ago, I first ran across the Space Men out there. (He waved his hand toward the west away from the sun.) But after I had studied them for a time, I knew that there was really nothing very remarkable or impossible about the nature of their living. It is actually quite similar to our own.

"Why do we need air? Simply because by the chemical combination of oxygen with food we obtain the energy necessary to make our brains to think, our limbs to move, and our hearts to beat. Energy is life. But doesn't it occur to you that this vital thing might be obtained in some other manner? The Space Men do. Their principal food is the radio-active element, atomic number 109, as yet undiscovered on the planets. It is a purplish liquid that is fairly abundant on a number of the planetoids. Daily, like radium, it gives off vast quantities of energy; and when in the systems of the Space Men it supplies them with power more efficiently than food and oxygen ever could do for us.

"Why can't we survive the intense cold of space? The answer is a simple one. The protoplasm of all forms of living things that we know of, including the Space Men themselves, is a colloidal jelly the principal portion of which is, and must be, a liquid. Cells must be bathed and nourished, and impurities washed away. Without liquids there seems to be no likelihood that there would be any life, unless in some manner a gas could perform this fluid function. Solids would remain forever dead and motionless.

"If anything happens to chill even slightly the protoplasm of any of the higher forms of planetary life, the body fluid becomes sluggish and death may result. No mammals or birds that we know of can live actively with their body temperatures at all approaching the freezing point of water. However, in the polar seas of both planets there are creatures whose systems function quite normally with their blood temperatures just above this point. But beyond this deadline, zero degrees Centigrade, or a little lower or higher, depending on the actual congealing point of the water in their bodies, even they cannot go, for there, the cold limit of Terrestro-Martian life has been reached.

"Why couldn't these polar fish survive the cold of space? Simply because the protoplasm of their tissues, based on water, would instantly become solid, and in solids as I have said, there can be no real life except perhaps in the form of suspended animation.

"The Space Men face no such danger, for first, their bodies are protected by this heat-resisting outer covering; and second, the liquid in their veins freezes only at absolute zero, and since it is radio-active—producing heat from within itself—it cannot get that cold even in the void. And that, friends, is the whole stupendous, simple explanation."

"And how do the Space Men's vehicles move?" asked Jan.

Hekki shook his head. "Except that a strange propulsive ray is involved, I know very little about it. I have not yet discovered how the Space Men manage to produce the ray. The works of Nature ever surpass the works of man.

"And that is all I have time for now, my friends. Breakfast is ready aboard ship. Enjoy my hospitality to the fullest!" Hekki's mask of smiling sardonic cruelty had dropped again. He waved something to Sega.

Janice, sensing that she was about to be separated from her lover, threw herself into his arms. The series of things she had gone through in the past twenty-four hours had frayed her nerves almost to the breaking point.

"Don't let them take me away from you, Austin. Don't let them! Oh, Hekki, please!"

Hekalu's face reddened, and then Sega tore the two apart. Shelby struggled but it was useless. Sega's huge muscles were quite equal to the task of mastering a dozen of the best fighting men of Earth.

He dragged his captives aboard the *Selba*, and guided by the inscrutable Koo Faya, locked them in chambers from which escape would now be definitely impossible. Jan was thrust into the room she had occupied before, but Shelby was put into a chamber somewhat larger than his original prison.

An almost ungovernable fury had taken possession of the young Earthman. If for only a moment he could get his hand on the smooth Hekalu! His fingers clutched and unclutched spasmodically as he hurriedly paced the room. When presently, he found himself hammering on the walls with the frenzy of a trapped gorilla, a realization of where he was headed came to him. "Stop where you are, you fool!" he muttered to himself.

He went to the table where an appetizing breakfast was set out. He ate a little and then waited a while. He wanted to make sure that the food was not drugged. Half an hour passed and he felt no ill effects. He ate the rest of his breakfast. Then he made several attempts to signal Jan by tapping on the walls, but he was quite sure that to get a message to her in this way was now out of the question.

For a long time he gazed out into the sunlit valley floor from his window. Preparations of some kind were under way. It looked as though the entire population, which must have numbered close to fifteen hundred Space Men all told, was getting ready to move away *en masse*. Scores of the strange black people were hurrying about, lugging loads of weapons and hundreds of large cylindrical objects into four immense box-like things of dull metal. Several vehicles, resembling machines of the Space Men, but many times larger, were clustered together in a group.

It must have been several hours after Shelby had been taken into the space ship that two of Alkebar's people came to his room, carrying between them the unconscious form of the Space Man who had been Jan's and his fellow prisoner during the night of their arrival on Mars. They threw the limp giant down carelessly on one of the bunks, and without a glance at him or the Earthman, they stamped out.

Shelby would have liked to examine his cell mate more closely, but owing to the chain which had again been fastened to his ankle, it was impossible to get nearer to him than four yards. Who was this creature? His gorgeously bejeweled harness and his huge size seemed to indicate that he had been a leader of some kind. Shelby had noticed that all Space Men who had a right to command, were somewhat larger than their fellows.

All through the long Martian day Shelby paced the length of his tether, pausing occasionally to look out of the window and to think. By nightfall he was in a state bordering upon complete dejection. Not that he was weak; Shelby could face trying situations shoulder to shoulder with the stubbornest and cleverest men that Earth or Mars could produce. But he was human and had his limitations. Recapture after a glowing promise of freedom and safety for his people, his love, and himself had almost crushed him.

Only half interestedly he wondered when Hekalu Selba would strike. He knew that it would be very soon. In vain he tried to tell himself that he had no real proof of the Martian's power, but always a vision of those black horrors swooping down like living thunderbolts upon Taboor or New York or Chicago made him realize how futile would be any resistance that the planets could offer.

Even if there were but fifteen hundred Space Men, and Shelby was certain as actual knowledge that there were many more, and even if they must fight with their bare hands, still they would be a formidable menace. Within an hour's time they could strike in a dozen different places on the surface of a planet. Shelby did not know that already there were forces of

Fate in action which neither he nor Hekalu Selba himself had been able to foresee — forces however, which boded no good for the worlds.

Koo Faya brought the Earthman his noonday and evening meal. With each came a note from Hekalu, both exactly alike: "Remember the Atomic Ray." Doubtless the Martian sought by endless repetition of this message to undermine his captive's nerves to a point where he would divulge the secret.

At dusk there was the sound of activity aboard the *Selba* — muffled shouts and the drone of generators being tuned up. Then the slow rocking and swaying of the vessel which told that her levitator plates were in action, raising her off the ground, through the atmosphere and out into the void.

Shelby looked out of the window, saw that the stars were growing brighter and the sky blacker. A searchlight was playing from somewhere on the ship, for in the shadow of the planet it was very dark. The beams swung back and forth stabbing through the swarms of Space Men who flew in a cluster about the *Selba*. The lights lingered for several instants on the forms of four great metal cubes that were being lifted up through the gaseous envelope of Mars by a number of the larger discs the Earthman had seen resting beside them in the valley that day.

Shelby threw himself upon his bunk. He gave one quick glance at the blob of darkness on the other bunk at the farther end of the room, wondered vaguely who or what the creature could be, and then, mentally and physically exhausted, went quickly to sleep.

When he awoke Shelby spent many minutes staring at his fellow prisoner. There were indications that his consciousness was returning for he stirred frequently. Presently he who had been the Earthman's and the mysterious one's jailer in the hut the night before, came, bearing a bowl filled with a purplish radio-active liquid which served the Space Men as food. He also carried a hypodermic syringe and a small glass container partially filled with a clear fluid.

These last two articles he placed upon the table, while he carried the bowl over to his charge. He shook the lacerated and bejeweled Space Man roughly and when he had aroused him to a sluggish half-consciousness, held the bowl of liquid food to his lips. Mechanically the prisoner drank.

Shelby looked at the tiny vial on the table and then at the back of the jailer. Close beside the vial stood a glass partially filled with water. The Earthman had drawn a drink from the tap shortly before going to bed, and had left the tumbler standing there.

The idea that had now entered his head had no real purpose. He recognized it as no more than a practical joke, plain and simple; but the idea was clamoring for attention. He would pour out the drug, which was almost certainly meant to keep the giant captive senseless, and replace it with harmless water. The jailer would not see for he was very busy. A little noise, the rattling of the chain or the tinkling of the glass as it was set down, would not matter, for though the Space Men may have possessed a very delicate touch sense capable of detecting faint vibrations in solid objects about them, Shelby knew by now that they had no real organs of hearing.

And so, quickly the deed was done, and quickly he returned to his bed feigning sleep.

It was a long time after the jailer had departed before Shelby's trick bore fruit. The huge prisoner rose to a sitting posture and looked about, a trifle dazedly at first. He surveyed his wounds, felt over himself tentatively, and then glanced at Shelby. The Earthman saw that the fogginess was clearing from his big eyes. There was a questioning expression in them.

Shelby thought that there was a slight chance that the colossus might be able to read his lips even though he could not hear. "Who are you?" he questioned in Pagari.

Apparently the creature understood, for immediately he turned, and with his forefinger slowly traced out on the wall behind him in the planetary symbols: "Friend of enemies of Black Emperor and of Man from Fourth World."

Shelby was taken aback by the Space Man's startling knowledge of things of which he should know nothing. "That makes me your friend," he wrote, smiling.

The giant nodded, and for almost a minute stared fixedly at the Earthman. There was a strange appeal in his eyes. Finally he turned, and laboriously he traced a quaintly worded message on the wall: "Think hard to know what I go say," he wrote.

Shelby had heard a good deal about telepathy and thought transference, depending on etheric vibrations of some kind, supposedly originating in the mind of one individual, and capable of being detected and interpreted by the mind of another. Several savants of Earth and Mars claimed to be adept with it, but owing to the fact that to master the art required a long period of intensive practice, it had not come into general use.

Could it be that this savage of the void was claiming knowledge of it? Sensing the meaning back of the odd words, the Earthman bent every fibre of his will to the task of concentrating on the idea of communication. He gazed fixedly at the eyes of the black mystic, and presently felt a slight tingling about his temples, and then, within his brain it seemed that a tiny voice speaking with a queer wording and a peculiar accent, came to life. It was odd to look at that blank impassive face and hear those words!

"I know you to be friend of mine," the voice said. "I read it in brains. You free me from sleep. But where are we? What Fourth World Man do? What for you here?"

Briefly Shelby outlined the events of the past few days, starting with his meeting with Hekalu. However, he was careful not to make any mention of the Atomic Ray. Then, partially through curiosity, and partially in the hope that the information might be helpful, he mentally asked his companion to tell him more about the Space Men's relations with the Martian.

"Everything maybe all right," said the giant. "Maybe everybody happy at last. Who know? But I tell you. We Star People—my people Star People. For a long time, oh, for very long time, we wander out there in empty places. One million year, two million year, who know? We free. Maybe find little planet—we camp there—soon go away. We fight, we hunt. Oh, there very many of us! Like sand in sky!

"One day some of us find your sun. We land on little world. Stay long. Man from Fourth World come in ship. We frightened, but he make friends. Bring us gifts. We give jewels and things we make. He learn our sign language—talk with us—tell about his world. Go away but soon come back. Bring more gifts—want more jewels and things. He take some of us with him to empty desert where nobody live. Tell us to bring jewels there to trade, but always be careful no one see!

"He make friends with Black Emperor. They plan. Gather big army. But many not like Black Emperor and Fourth World Man. My father, big noble, not like them; I not like them. They never good to us—make our people work hard, and take away our animals.

"Civil war soon—my father lead many little tribes, but Black Emperor and Man from Fourth World win. Have many strange weapons. Make peace for big conquest war, and I am hostage on Fourth Planet.

"Mars man good to me at first. I learn languages—both Pagari and Earth language. I learn to throw thoughts. My father learned from Mars slave. Then bad things happen. Fourth World Man not like me to throw thoughts to my father so far away. He give me sleep drug. When my father lead revolt again, Mars Man torture me. Now, as you say, he take me back to place where army is, on two little worlds."

A gleam of hope came into Austin Shelby's eyes, but it passed quickly. His lips curled bitterly. It was not well to base one's hope on the assertion of an unknown savage that he could hurl his thoughts across millions of miles of space.

"By what name are you known, Man of the Void?" he asked.

The voice in his brain spoke again: "Mars Man call me Ankova." Here the giant made a darting gesture with his hand. "Mean same as so in my sign language—Darting Meteor."

"I see. Can you communicate with your father now, Ankova?—get help?"

The Space Man nodded. "My brain clear now," he said. "Sleep drug not bother me any more. I talk right away."

CHAPTER VIII
THE BATTLE IN SPACE

He lay back on the bunk and for several minutes stared fixedly up at nothing. The performance was reminiscent of the seance of an ancient spirit meeting. He sat up, and again his big eyes fastened themselves upon Shelby, and the uncanny voice spoke in the Earthman's brain:

"I get father. He on scouting expedition—very close. He bring five thousand men to rescue you and me. They get here maybe three, four hours. My father—his army same weapons as Black Emperor's. Flash, flash—all gone—everything gone."

There was the sound of movement beyond the door. Shelby waved his hand in a quick downward gesture which Ankova interpreted correctly. He slumped limply upon the bedding in a very excellent counterfeit of unconsciousness. And then Hekalu Selba entered. His face was white as chalk, and yet there was nothing in it that hinted even of a trace of fear—only icy calm. Behind him was Sega.

"Mr. Shelby," the Martian said with slow cool deliberation, "think well. Either you will reveal the secret of the Atomic Ray immediately or I shall have you immersed in the juice of the flame flowers."

Austin Shelby met Hekalu's chilly stare with a taunting smile. He sensed in the Martian's manner that his plans had met with some serious danger.

"Though I am your prisoner," he told him, "I believe that I can defy you. In the first place I do not fear the tortures that you might inflict upon me." Here he took a tiny glass capsule from his sleeve pocket and placed it in his mouth. "I do not mean by that that I am super-human, that I can endure any pain. But should the torture become unbearable I would crunch the poison vial which I have carried since I joined the Sekor fraternity back on Mars, between my teeth and bring death. That, I am not afraid of. Besides, I could give you the formulas for almost any number of unknown compounds, any one of which might be the missing crystal for all you might know. It would be several hours before you would discover that I had not given you the right one."

The Martian's face grew even whiter and harder at these words. Thoughts and plans flashed through his mind. Should he tell the Earthman what had happened—that Alkebar, the Black Emperor, had secretly slipped through the air lock into space?—that he was certainly intent upon conquering the planets alone? It would not be hard to convince the Earthman that the savage Alkebar would be an infinitely more terrible and ruthless master than any human being ever could be. Perhaps he could win Shelby to his side for as long as he needed him. He was wavering, and then, with the sudden rush of inspiration a better idea came.

"I have told you many times that you are clever, my friend," he said with some slight show of his old careless air. "Again I compliment you. But listen carefully: suppose I took the girl—put her in the gentle embrace of the juice of the flame flowers—told you to produce a formula that would work before I released her?"

The effect on the Earthman was electrical, but it was not quite what Hekalu Selba had expected. The blood red haze of murder rushed before Austin Shelby's eyes, and with movements more suggestive of a wounded panther than a human being he leaped from the bunk and tore for the Martian with flailing fists. He gave no thought to the idea that what Hekki had said might be only a histrionic gesture.

"Oh, God!" he shrieked raspingly, "You Devil! You unutterable stinking, rotten fiend!" But it was a wild useless move. Hekalu was lightening quick and sure with the pistol. He inflicted death, or merely produced a disabling wound almost at will. And so it was that Shelby sprawled senseless on the floor with a nasty though not very dangerous bullet wound across the side of his head.

Sega and the Martian were bending over him, and then again the unexpected happened. An ebony form whose great hands and incredible muscles seemed quite equal to the task of tearing a gorilla limb from limb, arose from the other bunk and towered over the Prince of Selba and his Space Man companion.

The former, hearing a slight sound, turned, and realizing his peril fired two shots at the mountainous monster. Then he darted agilely for the door. He gave one quick backward look—saw the hand of Ankova descending with trip-hammer force upon the skull of Sega, and then slammed the stout portal behind him.

Sega had been unfortunate, but now all his troubles were over for his neck was broken. Ankova transferred to his own belt the weapons of the corpse—his heavy pistol—his case of atomic grenades—his bejeweled war club. Then he devoted his attention to Shelby.

Gently he carried him to the bunk and made awkward attempts to bandage his head with strips torn from the bedding. Satisfied at last with the crude but effective results of his efforts, he strode to the window.

For a long time he stood there, staring. But he saw nothing that interested him. The ether all about was crowded with Space Men coursing with the *Selba*. Except for a gentle swaying shifting movement they seemed to hang perfectly motionless in the void, and yet their speed was many miles a second.

The fantastic cavalcade aroused no wonder in the mind of Ankova, for to him they were as prosaic and commonplace as the grass under the feet of any Earthian. He cocked his head on one side as though listening. Perhaps at that moment something was coming to him from across the endless regions of the etheric desert — something which only his incredibly refined telepathic sense could detect.

His unshod feet sensed the faint vibration in the metal floor. Someone was approaching the room. First taking the precaution of tearing Shelby's chain from the wall, he turned and waited before the door with ready war club. He did not wait long for it banged open almost immediately. A Space Man appeared. Behind him were others.

Ankova did not ask their mission for he saw that they wore the insignia that meant loyalty to the man from the Fourth World. Instead he leaped in to close quarters. His whirring war club, toothed with sharp spikes, ripped and tore at the head and shoulders of the unfortunate warrior. Falteringly, the creature tried to parry the blows with his own weapon; but it was useless. Before he was able to attain his fighting stride he was down, the purple radio-active liquid that flowed in his veins in lieu of blood, dyeing the threshold. His lips curled in a grimace of agony, but he made no sound — mute he had lived and he died in the same manner.

Ankova stepped over the prostrate form and engaged the one who had stood behind him. The second Space Man fared little better. He made but a brief and unsuccessful defense and then he too went down. And so Ankova, who before his capture had won fame among the tribes of the Star People as one of the mightiest fighters that their race had ever produced, battled on in the narrow passage until the seven Space Men whom Hekalu had sent to put him and Austin Shelby under restraint were either dying or dead.

The victor glanced down the corridor — saw at the farther end a small portion of the control room's interior. Koo Faya, the Martian, was there, working with demoniac haste over switches and dials.

Ankova drew his pistol, started to aim at the slave, and then thought better of it. There was a tenseness within the hull of the *Selba* — something which made a deep impression on Ankova's keen intuition. His muscles tautened and a tingling sensation rippled over his ebony hide. The vibrations of the rocket motors were more noticeable than usual. Evidently the ship was tearing along at the greatest speed it could attain. And it swayed unnaturally.

Ankova knew the layout of the *Selba* well, for he had traveled in it often. And now he sensed quite clearly what was happening. He hurried to a supply room and selected a space armor from a rack. His Earthman friend might need it. Then he dashed back to the room In which he and Shelby had been imprisoned.

A glance out of the window confirmed his suspicions as to what was going on. The force of Space Men which was acting as an escort for the *Selba* had arranged itself in a sort of spherical protecting network around the craft. Another and superior force was attempting savagely to pierce this formation. The foes of Hekalu's henchmen would draw themselves into cone-shaped groups and rush the defenders, and the latter would swarm over the cones like angry and determined hornets. A hot fight was in progress out there. The ether was lit with green flashes of light, and fragments of the bodies of Space Men and their vehicles already strewed the void. In this running battle the *Selba* was not idle. Her torpedoes were exploding among the attackers with blinding glares of light.

Ankova wondered who the would-be destroyers of the *Selba* were. Clearly they were not the forces of his father, for they had not yet had time to arrive. Some stray tribe perhaps. He wished that he might see their insignia, but owing to their distance from the ship and their eccentric movements, this was impossible. He did not know that they were the minions of Alkebar who had turned enemy to Hekalu but a few hours before.

The Space Man realized that for the time being he was safe enough, but he took the precaution of planning for escape from the ship should it become necessary. He eyed the heavily glazed porthole. A few deft blows with his war club would shatter that. Beyond, there were a few discs without Space Men circling about. With luck it would be possible to capture one. First he barricaded the door with metal bars torn from the bunks, and then put the space armor on the still senseless Earthman. Then there was nothing to do but wait.

The battle was going against the defenders. Shattering concussions of atomic projectiles banging against the *Selba's* hull made the hurtling vessel pitch and roll frightfully. The thunder of shells waxed and waned.

It must have been over two hours later that a huge torpedo set in motion by the forces of the Black Emperor, struck the ship. The explosion rolled her completely over, and tore a jagged though not disabling hole in her side. The air puffed out from the control room compartment, but the men who labored so feverishly there, were clad in heavy space armor, and aside from being badly bruised they were unhurt.

The torpedo was the last gesture of the Alkebarians. Ankova saw a cloud of luminous specks approaching from the void at terrific velocity. They grew rapidly brighter. A blue and an orange star shot up from their midst—the identification signal of Telaba, Ankova's father. That signal was quite enough for the Black Emperor's men. Without waiting to argue they turned and fled. So quickly did they go that Telaba's warriors were unable to identify them.

The rebel tribesmen were checking their speed now, preparing to fight. But still they came on apparently like hurtling comets. They swept the remnants of Hekalu Selba's loyalists before them in one terrific charge, and then they were swarming over the *Selba* and through the rent in her side. There was a brief flurry of pistol shots from the crew before they were captured and bound.

In a prison compartment aft, Austin Shelby had regained his senses sufficiently to have a vague idea of what was going on around him. Ankova was supporting him, and he was staggering toward the door. His mind took up a train of thought from where it had left off. He was calling for Jan and cursing Hekalu. Cased as his head was, in an oxygen helmet, his shrieking voice was magnified a dozen times, and assumed a weird vaulted quality that startled him back to sensibility.

Ankova read his thoughts, and by telepathy replied to him: "Your lady? I forget. But we find her. She all right—sure!"

The Space Man removed the barricade and opened the door. The sudden outrush of air from the room almost toppled Shelby from his feet. And then the Earthman heard a familiar voice in the head-phones of the radio with which his helmet was equipped: "I'm in X7, Austin. Let me out if you can."

"Janice!" he cried, and with new vigor hurried to the door of the room she had mentioned.

Ankova smashed the lock with his war club and the portal flew open. Jan was standing there encased in space armor. She was trying hard to smile.

"You're safe, darling!" Shelby cried, "And I thought that that fiend was going to hurt you!"

"My luck," she said. "Koo Faya was thoughtful enough to bring this space armor, otherwise, I wouldn't have been fit to look at any more." She pointed to a shattered window. "And you—heaven's how you can yell—and swear! I am ashamed of you!"

Her eyes widened when she looked at Ankova, but Shelby reassured her. "This is Ankova, and he is our friend—big shot, too," he said. "And Jan, I guess we're free now—really free."

Ugly Space Men, some of them gashed and wounded, crowded about as though bent on destroying the two feeble Earthians. But with imperious gestures Ankova waved them back. He conversed by signs with these warriors of his father, and then took Janice Darell and Austin each by the arm.

"Big surprise," he told them. "Come."

He led them to the control room. And there, in the grip of a black colossus was Hekalu Selba—captive. The Martian nodded perfunctorily to the girl and then turned his level gaze toward the man. His face showed no hint of anger, and it seemed that a shadow of a smile twinkled about his lips.

"Here we have a contrast, Mr. Shelby," he said quietly, "triumph and disaster staring at each other!"

Shelby told him that he should be wreaking vengeance on the noble for the numerous wrongs he had done him, but the calm unflinching attitude of the Prince of Selba made him almost like the captive.

Shelby waved the Martian's captors back and he stood free. "There is no contrast now, Akar Hekalu, for an outsider could not tell which was which!"

As Hekki's jailer led him away, Shelby, assisted by Janice Darell, busied himself with the ship's controls.

And so the battered *Selba* escorted by five thousand Space Men set out for a certain minor planet where were amassed the forces of Telaba, insubordinate vassal of the Black Emperor. And on another planet was Alkebar, the Black Emperor himself, ready to hurl his shock troops, a horde five million strong, at the planets.

CHAPTER IX
THE REVOLT OF ALKEBAR

The light of a shrunken sun shone down coldly and ineffectually upon a jagged and distorted landscape. Along the horizon, which was strangely abrupt, twisted gray hills loomed up with harsh clearness against a black starlit sky. There was no atmosphere to soften their lines, nor to dull the needle-like points of deepest sable that were their shadows.

In the foreground, which was a fairly level plain, were hundreds of hemispherical shelters hastily built from loose fragments of rock. A vast horde of Space Men hemmed them in. The sunlight glistened on the ebony hides of the warriors and on their polished accouterments and weapons. Some of these rebels of the void were greedily drinking the purple radio-active liquid which meant life and strength to them, and attendants were hurrying about carrying large canisters of the food to each unit of Telaba's army. Most of the men crouched expectantly beside their discs, waiting.

In a small metal building, which the Man from the Fourth World had recently had constructed for his own use, four people were gathered. Two were Space Men, and two belonged to the green planet called Earth. One of the Space Men was talking, not with his mouth for he had no vocal cords, but by means of fine mental vibrations which caused a feeble high-pitched voice to speak within the minds of the Earthians.

"I owe you great debt of gratitude, Mr. Shelbee—you help to save my son from Alkebar and Fourth World Man. Telaba do not forget this. I do what I can. But that is little. Black Emperor start to smash Earth and Mars soon. Perhaps right now. Perhaps in hour. Who know? Spy send signal any time now. We outnumbered ten to one. Alkebar crush us, wipe us out like that!" He slapped his palms sharply together. "But we do what we can, Earthman."

Shelby took Telaba's cold hand for a brief hearty handshake. "Thanks, Telaba," he said simply. "Jan and I certainly appreciate what you are going to do for us and our people, and I know that if we are successful, the worlds shall be mighty grateful too. They have ways of showing their gratitude. But don't be so sure that we are going to fail. We have the *Selba*, you know, and a new weapon that has never before been used.

"Hekalu was good enough to construct an immense projector for us. Except for the resoldering of a few wires, and the insertion of a tiny but important crystal which I happen to be carrying with me, it was complete and ready for operation.

"The ship is fueled and ready for action at any moment. When the word comes and we set out, annoy the forces of Alkebar, but do not engage or mix with them any more than you have to. I'll be somewhere around, ready and glad to spray them."

"What do you mean, 'I'?" Jan put in. "It's 'we,' because I am going along!"

Shelby knew that the undertaking he had in mind was but an ace from certain death; but he did not argue with the girl. Her cool wit and nerve would be very helpful, and besides there was little choice, for death was grimly in pursuit of all of them.

"Right you are, soldier," he said laughingly. "My mistake!"

A red light bulb flashed on the wall, and then, without waiting for permission, a Space Man rushed into the room, his arms waving wildly, forming frantic signs of the Star People's deaf mute language. Bent in a half crouch, his great arms flexed, Ankova translated for the benefit of the Earthians:

"Fourth World Man escape—in *Selba*. We are betrayed—someone help him. He out of sight already. Going to help Black Emperor. And now red star burns in space—spy's warning—Alkebar forces start!"

Telaba rushed to a big lever and pulled it. Immediately a huge trip hammer began to pound ponderously on a metal plate set in the ground outside the building—sending vibrating pulsations out through the crust of the planetoid—the alarm signal which would be sensed by everyone of Telaba's men, telling them to be ready for instant action.

The four looked at one another. Each knew what this last move of the Prince of Selba meant, but no one thought for a moment of giving up the fight.

"It won't do any good to pursue the Martian," Shelby cried. "That ray projector of his—he'd blast us out of existence. All we can do is try to hinder Alkebar's invasion—seek to delay him. If I could only somehow get through to Mars with the secret of the Atomic Ray! Telaba, haven't you a ship capable of carrying a large enough oxygen supply to last me for the journey?"

"Never mind!" Ankova cut in. "I go! Many times I been to Mars. Give me plans. I go right away. I get them to fight."

Shelby drew from his sleeve pocket the black case containing information concerning the Atomic Ray which he had recovered from Hekalu Selba at the time of the Martian's capture. He opened it, and with his stylus added a brief message to the mass of notes inside, and wrote down the formula for a certain complex chemical compound. Then he handed the case to the Space Man.

"Take it to Alman Mak in the Checkald of Taboor if you can, Ankova. Good luck."

The son of the rebel chief hurried from the room with the missive in his hand. Shelby knew in his heart that to attempt to get Earth and Mars into action in time was a useless gesture, but he could not suppress a thrill of admiration for this wild son of the void. There was hard mettle in Ankova's makeup, hard and true. And most of them were like that—most of Telaba's men anyway.

"You two come with me," Telaba was saying. "We fight together. Put on space suits." He was tapping an instrument resembling a telegraph key. In unison with his movements the heavy signaling hammer sounded out orders and commands to his forces.

When the Earthians had eased themselves into their heavy protecting attire, Telaba led the way down a spiral stair and through an air lock, out into the open. Here everything was grim silent activity. Group after group of mounted Space Men poured skyward. Telaba's army was a mighty thing; with luck it might have beat down the resistance of either one of the two planets. But when compared with Alkebar's colossal horde, it paled into pitiful insignificance.

Nearby, a space disc, which must have measured fully two hundred feet in diameter, rested. The three mounted the light ladder which led to the interior.

In the metal walls were mounted two heat-ray projectors of Martian design, as well as several torpedo catapults and machine guns. Two Space Men were inspecting them.

Telaba signaled to the driver who knelt with lever in hand. The great disc trembled and the propelling force which no human being had yet learned how to produce, sent it and its burden hurtling toward the stars. The minions of the rebel chief circled and swirled about their commander's ship in wild soundless salute.

Telaba was operating the signaling mechanism which fired lights of various colors up through the roof of the armored coach, and in reply to his flashing commands, his horde formed a monster cone which shot with ever increasing speed through the void.

A sickening giddiness came over the two Earthians, for there were no devices to produce artificial gravity here. It was the space nausea which had made early interplanetary travel such a nightmare. The Star People, born where gravity is almost unknown, were of course not affected in the least.

Clinging to stanchions and hand grips to keep themselves from floating free, Janice Darell and Austin crept about the floor examining the weapons and scanning space ahead for signs of the enemy. They disliked to admit to each other that they were very sick; but if they thought that it was possible to forget the retching pains in their stomachs by diligent devotion to other things, they were mistaken.

Their suffering continued until Jan remembered that the force of this almost forgotten malady could be reduced by lessening the amount of oxygen taken into the lungs. A few turns of the intake valves of their helmets accomplished this, and they soon felt much better.

It was a long time before there were any indications of the near presence of the enemy. Ahead, two asteroids glowed, a dull red. One was quite close; the other farther away. It was Shelby, peering steadily through his binoculars, who first discovered the glowing cloud, thin and faint like the nebulous substance of the Milky Way, pouring up like ghosts' hair from the rounded pate of the nearer asteroid. He knew that it was made up of countless points of light, too small to be detected individually. Not long afterward Telaba discovered a similar cloud coming from the second of the minor planets.

The rebel chief's greatest advantage, if he had any at all, was that of surprise. Because of its comparatively small size his force had probably not yet been discovered by the enemy.

Coolly he flashed the order for long-range bombardment formation. Instantly the army spread out, forming a thin rectangle whose broadest surface was perpendicular to the line of firing between the opposing hordes.

A second or two later the first rocket torpedoes of the rebels went, spewing fire, toward their goal. In a steady swarm others followed them. The missiles were not radio controlled and fitted with tiny television apparatus as were a few of the torpedoes employed by the Interplanetary Traffic Lane Patrol, but since the approximate range was known, it was easy to set the time fuses so that the atomic charges would explode in the midst of the densely-packed enemy.

Without asking anyone's permission, the Earthians had appropriated a pair of catapults and were working them like demons. As fast as they could cram the ten-pound rockets into the breeches of the tubes, the projectiles streaked out in flashes of green flame toward the nearest of the nebulous clouds.

Shelby was sweating furiously from the exertion, and the moisture absorption apparatus of his space armor was putting in some tough service.

Occasionally he glanced at Janice working beside him. Her face, visible through the glazed front of her helmet, was white and set—almost hard. And there was boundless determination in the firm curve of her little rounded chin. He liked her attitude, but it was better to take it easy until the real fighting began.

"Slow up a bit, soldier," he remarked into his transmitter. "Powder your nose!"

Her face brightened as she turned toward him. "I wish I could powder my nose," she said, pouting. "Only I can't reach it!"

"Too bad. These space suits rob a girl of so many of her exquisite little tricks."

"Well," she put in, "I can still cover up my yawns with my hand if I find this pastime too much of a bore." They both chuckled at this little joke.

Janice took the last missile from the case she had been emptying and rammed it home. She jerked the lanyard, and with a thudding jolt the torpedo was on its way. Then she paused to scan the horde of Alkebar through an observation port. "Hurrah," she cried, "we're scoring!"

Without discontinuing his hurried feeding of his smoldering piece, Shelby looked up. The cloud had grown considerably in the few moments of action. It had cleared the asteroid now, and the other nebulous spot that marked the position of the Black Emperor's second army, was coming up to merge with it. In the midst of the first cloud, hundreds of minute specks of light were flashing—the atomic torpedoes were exploding. The sight reminded Shelby of what he had so often seen through the lens of a spintharoscope.

Alkebar's army continued to increase rapidly in apparent size. It looked like a monster amœba. But now the amœba was beginning to writhe, to swell up and grow dimmer. It shot out long sinuous pseudopods that seemed to grope angrily. Both Earthians sensed that the fight was about to begin in earnest.

With renewed vigor they fell to the task of loading and discharging the catapults; and close beside them the two Space Men who acted as gunners, labored coolly and methodically over their weapons, but with even greater efficiency, for their training had been long and thorough.

Telaba worked the levers of the signaling mechanism, and a brilliant purple star visible to all his henchmen shot up over the back of his beast. They saw it and read its meaning. Spread out to avoid enemy fire! As one man they obeyed, but they were none too soon. With abrupt suddenness the maelstrom of silent flashing death was upon them.

It was a pretty sight to the Earthians—those soundless globes of green flame that glowed dazzling for an infinitesimal instant, on the rich jewels and polished rifle barrels of the hordesmen coursing close by. But they were not deceived.

A Space Man vanished, torn to tiny fragments that mixed with the cosmic dust of the void. A huge disc, bearing a cylindrical battle car, was hit, and a jagged hole torn in its side. It twisted crazily, turning over and over. Austin and Jan felt the vibration of shell fragments banging violently against their own vehicle.

The nearer nebulous cloud had ceased to be a cloud now. It had resolved itself into a myriad swarm of dim specks which the Earthians knew were Space Men. Plainly Alkebar's minions were charging rapidly, bent on wiping Telaba's smaller force out of existence at one blow.

The bombardment doubled, tripled, quadrupled in intensity until it seemed that all space had turned to fire. Before the withering blast the army of the rebel chief was speedily being dissolved into drifting wreckage.

An exploding torpedo ripped several yards of armor from one side of Telaba's vehicle and reduced one of his black gunners to a mangled pulp from which the purple fluid spurted.

The force of the concussion turned the great disc completely over. Battered and blinded by the green glare, which exceeded even the sun of the void in intensity the Earthians tumbled against their weapons. Janice Darell started to scream but managed to check it—biting her lips savagely.

An explosive rifle bullet struck the huge vehicle, and it wavered.

Shelby spoke to Telaba who was clinging firmly to a stanchion with one hand and operating his signaling machine with the other. "Turn back, chief," the Earthman advised. "Our only motive is to annoy them and delay them. To continue this charge can mean nothing but destruction for our entire force."

Telaba sensed the mental vibrations that went with Shelby's words. "To turn back cannot do, Earthman," he said. And it seemed to the young engineer that there was a vibrant note of sadness in his telepathic voice. "Look! You see all guns and catapults point forward only. Not swing to rear—same on all gun cars. If run, not possible to shoot at chasing enemy.

Then they get us. That Alkebar's idea so his men must take offensive or die. He think that make them strong."

"But the riflemen are not so handicapped," Shelby persisted. "We can die here if necessary, but someone must live to carry on. Order them back!"

The chieftain shook his bulbous head. "To try what you say — useless. They not desert comrades or king. If I command, they disobey." There was a finality in his words which neither of the Earthians tried to dispute.

So that was it! Well, there was no sense wasting time talking. Shelby gripped a machine gun and sent a spray of explosive bullets ripping out into the ether. Janice did likewise.

As they worked their weapons they spoke rapidly to each other. "You understood what Telaba said? You know what that means?" Shelby asked.

"Yes. It's about the end of our tape, but that's nothing. We've been fairly lucky. All we can do now is hope that Ankova wins through to Mars in time, and fight like — like — "

"Hell!" Shelby's words slipped between clenched teeth, and Jan flashed him a quick smile even as their tracer streams crossed in the midst of a group of hurtling Alkebarians who had pressed too close together.

"Anyway, good luck!"

"And the very best of luck to you!"

The opposing forces were very close together now. The first of the Alkebarians were plainly visible — their long guns flashing — their ebony arms waving signals which probably passed for shouts of triumph among their ranks.

CHAPTER X
THE COMING OF THE ATOMIC RAY

Both armies had cut down their velocity enormously, but still they tore along at breakneck speed. And they moved like true Cossacks of the void, directing their machines by deft motions on the mysterious levers. Now diving, now climbing, now swinging this way and that to avoid the missiles of their opponents, they tore on. And death was everywhere.

No torpedoes were flying now, but machine guns and rifles were working terrible havoc. And so the horde of Alkebar closed with the forces of the rebel chieftain.

The machine which bore Telaba, directed by its skillful driver, dived and swung and zigzagged like a mad thing; but still the bullets rattled against the metal armor of the car. Its sides had been repeatedly struck, yet owing to its tough shell, had not yet been disabled.

Everywhere about it, mounted horrors whirled in an inextricable tangle, shooting and loading, and dying by the green flashes, their vitals strewing the ether.

Telaba had deserted his post at the signaling machine, for further orders were useless. For his rebels at least, it was every man for himself. He too was operating a machine gun.

The stars spun dizzily about the Earthians, as the machine beneath them careened in its insane fight. Every time a Space Man wearing a red circle on his breast crossed their sights, a burst spat from their hot weapons, frequently with good results.

A group of at least twenty Alkebarians sought to attack from the blind spot at the rear. But the driver twisted levers with a quick jerk, and the luckless riflemen found themselves facing four streams of steel. Those that could, darted out of range and renewed the attack from a different angle.

Frequently, throughout the battle, Shelby had wondered what had happened to Hekalu Selba and the Atomic Ray. Why wasn't he on hand to assist his ally, the Black Emperor? Oh, well, regardless of whether the

Martian was there or not the outcome would evidently be the same—only now it would be more dragged out.

The Earthian was surprised therefore, when suddenly the efforts of the enemy to exterminate them, which had been so intense in the brief moments since they had closed, suddenly lessened. Alkebarians were darting hastily toward the rear. Their actions did not suggest flight; it seemed that they were going to meet a new and more terrible enemy. The rebels could wait.

And the people of the rebel chief for the moment did not pursue—did not even fire. For they too saw! To the rear, in the center of Alkebar's horde, came the dazzling flares of explosions. So many and so close together were they, that they looked like a titanic conflagration of green flame. Against the light, the silhouettes of confused and bewildered space riders careened, like frightened pollywogs. The holocaust moved—swung. It was like a tapered column of fire veiled by a faint bluish haze.

The Earthians, Telaba, and the two remaining Space Men, forgetful of everything else, were staring in awed wonder at the phenomenon through the forward observation bay. It was Shelby who found the first part of the explanation.

"It's the Atomic Ray!" he almost shrieked. "Freeing the atomic energy in the materials that make up the bodies of Alkebar's men—literally causing their flesh and bones to explode! But how—what the devil—!"

"Look!" cried Jan. She pointed far up over their heads to where the cone of faintly bluish light swung, free from the milling horde. Up and up to its apex, and there hung what appeared to be a tiny cocoon of burnished silver.

The girl peered through her binoculars for a long moment. "I see the name. It is the *Selba*," she said. "Hekalu has made a mistake—he's attacking the wrong force! Or—or some ally of ours has gained control of the ship!" she hazarded.

"No time to make guess now," said Telaba. "To fight, much better." He had returned to the signaling mechanism, and was working it with cool efficiency, rallying his battered forces.

Like tigers they fell upon the Alkebarians, shattering them out of existence with a steady storm of rifle bullets. They met with only a weak resistance for the foe seemed to realize that the fates had played them false. The blue ray had been their promise, and now, like the sword of their ancient god of destruction, it was weaving calmly this way and that, snuffing them into nothingness. The Black Emperor's horde was dissolving, scattering.

Battalions of terrified Space Men poured past the rebel chieftain's car, shooting only hurried and ineffective volleys at their enemies, who pressed fiercely upon them. And never did Jan and Shelby miss a chance to spray them with searing bursts of machine-gun fire.

There was a lull. The Earthians took the opportunity to look up at the angel of death that was the *Selba,* far above. Most of Alkebar's huge army had already perished, or had dispersed in flight into the desert of space from which it had been recruited. But that the space ship would presently be engaged in a serious fight was evident.

A determined force which must have numbered a hundred thousand, was hurtling up at it, surrounding the craft with a halo of bursting torpedoes. At the head of the body of Space Men was a huge beast bearing on its back a car similar to Telaba's. Veri-colored signal stars spurted from it. Alkebar himself must be in it directing operations!

Coolly the guiding hand aboard the *Selba* was swinging his dreadful weapon this way and that, annihilating the attackers as one might annihilate a swarm of mosquitoes with a blowtorch. Half of them had already been reduced to those basic, intangible vibrations which constitute all substance. It was terrible, it was glorious; but what could it all mean? Hekalu's ship!

The still formidable remnants of the vengeance squadron was seeking to close in — to grapple with the vessel. The *Selba* was trying to dart out of their way, but the speed of the Space Men, a gift of Nature, was greater than that of this fastest ship designed by man. Grimly, in the face of almost certain death, they kept on. A score or so succeeded in landing on the curving hull, and, like leeches they clung to it. The Atomic Ray arched angrily, cutting a deep swath through those who still sought a hold.

And then the gleaming form of the *Selba* was completely hidden by the swarm of enraged horrors that poured over it. The Atomic Ray was snuffed out. The beholders saw the air lock being pried open, and the Space Men crowding into the interior of the craft. For a second the *Selba* wobbled crazily, and then her rocket motors ceased to flame.

"What are we waiting for? We have friends up there!" Jan cried.

Telaba flashed his orders, and the entire cavalcade charged toward the vessel, their guns spewing flame.

It was only a matter of a minute or so before that hurtling torrent of rebels had swept the Alkebarians from their prey. Those of the Black Emperor's men who had forced their way into the ship managed to hold the entrance for a short time, but under the urgings of their intrepid chief, the zealous rebels shot and hewed their enemies down as though they had been paper marionettes. The way was clear.

Telaba waved an order to his driver, and the space beast drew up alongside the *Selba*. Expectantly eager, the Earthians clambered aboard, followed by the chief.

The ship was a shambles. Its corridors were littered with bodies of Space Men who wore on their breasts the red circle which signified loyalty to the Black Emperor. Telaba's followers had done well.

The three made their way to the control room. Intuitively they had sensed what they would find there, and so, they were not surprised at what they saw — wreckage and the carcasses of Alkebar's warriors. The Martian had put up a stiff fight.

Shelby bent over the armored form of Akar Hekalu Selba which was sprawling on the floor, beside the pilot seat. A gaping hole in the tough metal plating under his right arm, and a thin trickle of blood, told clearly what had happened. "They got him," the Earthman muttered. "But why?"

Jan's eyes had wandered to the narrow desk before the pilot seat. There were the instruments and devices by means of which the ship was controlled, and there was the lever which had moved the ray projector in its mounting just beneath the nose of the craft. A calculating pad and a stylus were lying on the desk.

Something was written on the pad — a message. She called to Shelby, and together they read the brief, hastily scrawled note. It was in English:

"To Janice Darell and Austin Shelby, Greeting. Alkebar is breaking into the ship, and Telaba is coming. You will be with him, I know. From among my enemies I have chosen my friends. A man must have friends, and traitors do not serve. Forgive me for stealing your glory, Mr. Shelby. I shall be grateful. *Sidi Yadi*, Hekalu Selba, Akar."

Shelby looked at Jan and then at Telaba who was standing close beside them. "So that's it," he said slowly. "Nobody is totally bad."

"Not even Hekki," Jan put in. A hint of a wistful smile flickered about her lips. "I guess it's the end now," she went on. "A glorious adventure. Back to Earth!" Her voice had taken on a dreamy exultant quality.

"The end, Jan?" Austin asked. "Haven't you forgotten something?"

She looked puzzled, and then she laughed a brief gay little laugh which made roguish dimples twinkle in her cheeks. Even her fantastic attire could not hide her beauty. "You ridiculous old dumb-bell! Of course it isn't the end — just the beginning — with you!"

It was a considerable time before Shelby was able to repair the *Selba* sufficiently so that she could get underway for Mars but the task was finished at last. Escorted by the rebel chief's fierce hordesmen, they set out for the Red Planet.

Somehow, snatches of the ancient Bedouin song tinkled in Shelby's mind. He had read old books. "Across the desert I come to thee, On a stallion shod with fire...."

That did not quite fit the situation, for Jan was with him. But his steed, the *Selba*, was truly shod with fire. The rocket nozzles—and damaged though she was, she behaved like a thoroughbred. And out there in the void beside the ship—what were those shapes?—bizarre, impossible, yet real—real.

In docks scattered over Earth and Mars, battleships of space and their crews wait expectantly for an alarm that may never come. Telescopes comb the sky. Out there the Star People, new arrivals in the solar system, are shifting, moving about restlessly. But the planets feel secure. Their fleets could cope with the Space Men, were they a hundred times more numerous. And once in a while, on the desolate Sahara, or Mohave or Taraal, shadows come, settling down like flecks of darkness from the midnight heaven. They are Telaba's and Ankova's people. For a while—a day perhaps—they stay, bartering their exotic treasures for human wares. Then silently, mysteriously, they are gone, into the night....

www.ingramcontent.com/pod-product-compliance
Lightning Source LLC
LaVergne TN
LVHW041747190726
843493LV00008B/2493